TRACTION DRIVES

MECHANICAL ENGINEERING

A Series of Textbooks and Reference Books

EDITORS

L. L. FAULKNER
*Department of Mechanical Engineering
The Ohio State University
Columbus, Ohio*

S. B. MENKES
*Department of Mechanical Engineering
The City College of the
City University of New York
New York, New York*

1. Spring Designer's Handbook, *by Harold Carlson*
2. Computer-Aided Graphics and Design, *by Daniel L. Ryan*
3. Lubrication Fundamentals, *by J. George Wills*
4. Solar Engineering for Domestic Buildings, *by William A. Himmelman*
5. Applied Engineering Mechanics: Statics and Dynamics, *by G. Boothroyd and C. Poli*
6. Centrifugal Pump Clinic, *by Igor J. Karassik*
7. Computer-Aided Kinetics for Machine Design, *by Daniel L. Ryan*
8. Plastics Products Design Handbook, Part A: Materials and Components, *edited by Edward Miller*
9. Turbomachinery: Basic Theory and Applications, *by Earl Logan, Jr.*
10. Vibrations of Shells and Plates, *by Werner Soedel*
11. Flat and Corrugated Diaphragm Design Handbook, *by Mario Di Giovanni*
12. Practical Stress Analysis in Engineering Design, *by Alexander Blake*
13. An Introduction to the Design and Behavior of Bolted Joints, *by John H. Bickford*
14. Optimal Engineering Design: Principles and Applications, *by James N. Siddall*
15. Spring Manufacturing Handbook, *by Harold Carlson*

16. Industrial Noise Control: Fundamentals and Applications, *by Lewis H. Bell*
17. Gears and Their Vibration: A Basic Approach to Understanding Gear Noise, *by J. Derek Smith*
18. Chains for Power Transmission and Material Handling: Design and Applications Handbook, *by the American Chain Association*
19. Corrosion and Corrosion Protection Handbook, *edited by Philip A. Schweitzer*
20. Gear Drive Systems: Design and Application, *by Peter Lynwander*
21. Controlling In-Plant Airborne Contaminants: Systems Design and Calculations, *by John D. Constance*
22. CAD/CAM Systems Planning and Implementation, *by Charles S. Knox*
23. Probabilistic Engineering Design: Principles and Applications, *by James N. Siddall*
24. Traction Drives: Selection and Application, *by Frederick W. Heilich III and Eugene E. Shube*

OTHER VOLUMES IN PREPARATION

TRACTION DRIVES
Selection and Application

FREDERICK W. HEILICH III
EUGENE E. SHUBE
*Plessey Dynamics Corporation
Hillside, New Jersey*

MARCEL DEKKER, INC.   New York and Basel

Library of Congress Cataloging in Publication Data

Heilich, Frederick W.
  Traction drives.

  (Mechanical engineering; 24)
  Bibliography: p.
  Includes index.
  1. Traction drives. I. Shube, Eugene E., [date].
II. Title. III. Series.
TJ1095.H44 1983   621.8'5   83-13508
ISBN 0-8247-7018-8

COPYRIGHT © 1983 by MARCEL DEKKER, INC. ALL RIGHTS RESERVED

Neither this book nor any part may be reproduced or transmitted in any form or by any means, electronic or mechanical, including photocopying, microfilming, and recording, or by any information storage and retrieval system, without permission in writing from the publisher.

MARCEL DEKKER, INC.
270 Madison Avenue, New York, New York 10016

Current printing (last digit):
10 9 8 7 6 5 4 3 2 1

PRINTED IN THE UNITED STATES OF AMERICA

PREFACE

Traction Drives: Selection and Application, as the title implies, was written for the practical applications engineer who selects power transmission equipment. It treats the drive in its proper role as a major element in a system that consists of the power source, the drive, and the load.

Drives can be selected on a brute-force basis. For that, one merely chooses a large enough unit. With the price of a 20-horsepower drive about $6000 and of a 30-horsepower drive about $10,000, and with the cost of money high, there is a powerful incentive to understand the application in substantial depth so as to be able to select an optimum arrangement. The optimizing process is interactive and sometimes affects the choice of power source, the method of coupling, and even the driven machine itself. The total interactive process is highlighted in Chapter 12.

To do a proper job, the practical applications engineer needs to understand simple basic relationships, the principles of traction, the nature of traction fluids, the characteristics of power sources, the characteristics of loads, the method of optimizing the selection of the drive, and some practical economics.

The information required is available from diverse sources in several different countries and has heretofore never been assembled in one volume. This book has gathered a wealth of tabulated information about

Drives available from 30 sources
Standard NEMA electric motors
Standard IEC electric motors
Properties of materials
Conversion factors

Stress/life relationships
Characteristics of loads
Interest and annuity tables

which should prove useful to designers, students, mechanical and electrical engineers, and managers in many technical activities.

Many people helped us to shape this book and to assemble the technical information. Special thanks are given to Robert Carson, who was involved in many of the automotive developments. A prolific writer on traction, he contributed much of the information that comprises Chapter 2. It was Mr. Carson who first awakened us to the opportunities in industrial traction drives.

Charles E. Kraus submitted the entire text of Chapters 5 and 6, which comprises material updated from his earlier privately published *Rolling Traction Analysis and Design*.

We wish to thank C. A. Cass and R. E. Manville of Westinghouse Electric Corporation for the considerable support of their company in furnishing information on ac motors.

Special thanks are also given to Stuart Loewenthal of NASA, who offered direction in the area of stress and life. We thank Jean E. Kopp for his contribution of so many of the interesting traction applications. Geoff McLeod of F.U. (London) Ltd. submitted the analysis of the high-speed blanking press, which became the keystone of Chapter 12, in a way that enabled us to provide a focus for the entire book. Cam Tibbals of TEK Products contributed the hosiery knitting machine application. Gene Danowski of Sumitomo Machinery Corporation of America contributed the information on the paint mixing application. John Fischer of Koppers Company, Sprout Waldron Division, contributed the municipal waste compactor.

John Dixon, European sales manager of Plessey Dynamics Corp., was especially helpful in assembling the European standards documents and reference data that are included in Chapter 8. Ken Slezak of Plessey Dynamics Corp. set us straight on many accounting and economics matters.

Two ladies named Judy and Lillian tolerated a year-long mess and gave all the necessary encouragement while their husbands learned to type. For this we thank them more than anybody.

Frederick W. Heilich III
Eugene E. Shube

CONTENTS

Preface		iii
1	Introduction	1
2	History	5
	2.1 Introduction	5
	2.2 Automotive History	5
	2.2.1 Early Developments	6
	2.2.2 Fluid Development	19
	2.2.3 Modern Developments	19
	2.3 Modern Industrial Developments	24
	References	27
3	Generic Types	28
	3.1 Types of Traction Drives	28
	3.1.1 Kopp Variator	28
	3.1.2 Ball Disc/Roller Disc	41
	3.1.3 Ring-Cone	46
	3.1.4 Free-Ball	68
	3.1.5 Disc	76
	3.1.6 Toroidal	86
	3.1.7 Planetary	94
	3.1.8 Spool	106
	3.1.9 Offset Sphere	109
	3.2 Friction Drives	109
	References	115
	Appendix: Traction Drive Characteristics	116

4	Basics and Simple Formulas		120
	4.1	Fundamental Units: Customary System	120
	4.2	Derived Units	122
		4.2.1 Torque	122
		4.2.2 Work	123
		4.2.3 Power	124
		4.2.4 Energy	124
		4.2.5 Inertia	127
		4.2.6 Acceleration	129
		4.2.7 Efficiency	130
		4.2.8 Recirculating Power	131
		4.2.9 Viscosity	135
	4.3	Metric System	138
		4.3.1 Force	139
		4.3.2 Torque	139
		4.3.3 Pressure	139
		4.3.4 Power	140
		4.3.5 Temperature	140
		4.3.6 Summary of Metric Conversions	143
	Reference		143
5	Traction		144
	5.1	Traction Fundamentals	144
	5.2	Traction Coefficient	146
	5.3	Traction Fluids	154
	5.4	Spin	159
	5.5	Energy Factor	165
	Reference		167
6	Contact Area Analysis		168
7	Stress and Life		177
	7.1	Introduction	177
	7.2	Hertz Stresses	177
		7.2.1 General Case	178
		7.2.2 Specific Cases	178
	7.3	Life	186
		7.3.1 Life-Prediction Method for Designers	186
		7.3.2 Life as a Function of Load Spectrum	193
		7.3.3 Relationship Between Time and Percent Survival	196
		7.3.4 User's Method	199
	References		205

Contents

8	**Characteristics of Power Sources**	206
	8.1 Introduction	206
	8.2 Electric Motors	206
	8.2.1 Frame Sizes	207
	8.2.2 Standard Motor Horsepowers	207
	8.2.3 Mounting	209
	8.2.4 Enclosures	214
	8.2.5 Torque-Speed Characteristics	218
	References	234
9	**Characteristics of Loads**	235
	9.1 Introduction	235
	9.2 Torque-Speed	235
	9.3 Load Direction	236
	9.4 Drive Direction	237
	9.5 Cyclical Torque Variations	237
	9.6 Inertia	240
	9.7 Characteristics of Loads	241
10	**The Drive Specification Sheet**	246
11	**Economics**	261
	11.1 Capital Equipment	261
	11.1.1 Basis	261
	11.2 Depreciation	261
	11.2.1 Straight-Line Depreciation	262
	11.3 Expense	263
	11.4 Discounted Cash Flow Analysis	263
	11.5 Evaluation of Alternatives	264
	11.5.1 Net Present Value Acceptance Criteria: Opportunity Cost	271
	11.6 Interest Rate Tables	272
	11.6.1 Compound Interest	272
	11.6.2 Annuities	273
	11.6.3 Equivalent Interest Rate	274
	Reference	274
12	**Drive Selection: Examples and Case Histories**	275
	12.1 Introduction	275
	12.2 Traction Drive for a High-Speed Blanking Press	276
	12.3 Drive for a Municipal Waste Compactor	290
	12.4 Hosiery Knitting Machine	296
	12.5 100-Horsepower Paint Mixer	298

12.6	Cutoff Machine	303
12.7	Alternator Drive for Solar Energy Plant	307
12.8	Machine Tools	309
12.9	Other Industrial Applications	312
12.10	Additional Applications	316
References		321

Appendix A: Traction Drive Sources   323

Appendix B: Bibliography   329

Index   341

1
INTRODUCTION

Modern traction drives transfer mechanical shaft power from a source to a load efficiently by means of metal rollers running on a film of traction fluid against mating metal rollers. The rollers can be cones, cylinders, discs, rings, spheres, or toroids. The metal can be hardened steel, or chromium plating on some other metal. The fluid could be oil, silicone, or one of the new synthetic napthenics. It does not matter just what form each element takes, but three elements must all be present if the drive of which they are part is to be called a traction drive: (1) the input metal rollers, (2) the traction fluid, and (3) the output metal rollers.

This definition avoids the ordinary everyday meaning of the word "traction" (to pull). An automobile drive wheel on the road is not a traction drive. The wheel is rubber. The road surface is not a roller. The contact area is not lubricated. Many fine drives are excluded from this book because they use rubber belts operating against steel cones or because they use no lubricant in the contact area.

All of this is not mere semanticism. A traction drive is one that depends on low-rate shearing of a very thin fluid film to transmit substantial forces from its input members to its output members. The fluids used have been selected because they have a very high viscosity pressure index. When the local pressure on the fluid trapped in a contact area is momentarily increased to several thousand atmospheres, the local fluid viscosity increases enough so that the film supports a substantial normal force without permitting the metal parts to come in contact with each other. Only a low relative shear rate, on the order of 1 or 2% of the tangential velocity of the rollers, is necessary to develop the tangential drive forces required.

This is quite different from what occurs in a well-designed sleeve bearing. A sleeve bearing must be able to sustain a very high fluid film shear rate while exerting as small a tangential force as possible.

If a sleeve bearing operated in the same way as a traction drive contact, the tangential force exerted on its shaft would be so great it would be difficult to turn, and the bearing design might be regarded as a failure. If a traction drive roller spun freely against its mating roller (as any good sleeve bearing should), its design might be regarded as a failure due to gross slip. Yet the differences between good sleeve bearing design and good traction contact design are only those of materials and proportions.

As the definition implies, traction drives have to transmit mechanical shaft power from an input shaft to an output shaft. The drive would only be useful if it had a built-in speed ratio. Otherwise, a simple coupling would have sufficed to connect the input to the output. This ratio does not have to be variable to enable the drive to accomplish useful purposes.

The charm of traction drives is that the elements may be so chosen as to make the drive ratio

$$r = \frac{\text{input speed, rpm}}{\text{output speed, rpm}}$$

a variable under the control of the operator.

We recognize about 20 different ways of accomplishing variable ratio in a traction drive. In the familiar ball variator, a steel ball, which spins on an axle, contacts symmetrical input and output cones at different radii from the ball spin axis, depending on the shaft angle of the ball compared to the cone axes. For a given input speed the output speed may be varied over a ratio of 6:1 or even 9:1 when the maximum possible output speed is divided by the minimum possible output speed. Chapter 3 describes the characteristics of over 30 different drives and explains how the ratio change is accomplished for each.

Variable ratio has provided a powerful impetus for the development of automotive applications. The search for the continuously variable transmission (CVT) has led to increasingly better materials and lubricants and to increasingly competent designs. None has yet proved practical for the particular requirements of automotive transmissions, but active development work is in process in many laboratories. For afficionados of the horseless carriage, Chapter 2 should prove interesting.

This book is focused on industrial traction drives. These are rugged, quiet, well-designed, and well-manufactured drives which couple fixed-speed power sources to variable-speed loads—or variable-speed power sources to fixed-speed loads—day in and day out without attention or complaint.

Introduction

The environment in which these drives are used is often a competitive one. There are always several good ways of coupling a power source to a load, including some that require no separately identifiable transmission at all.

The principal competitor of the small (under 40 hp) traction drive is the variable-speed electric motor. This type of motor is coupled directly to the load, and its speed is regulated by an electronic controller. This competitor is not available in larger sizes because it is dependent on transistors. Transistor development has proceeded rapidly, but the larger sizes required for high-power usage are not yet economically feasible. As a result of this competitive pressure at the small end, the makers of the larger traction drives (up to 200 hp) seem to be enjoying better business growth than are makers of smaller drives only.

Traction drives find applications in places where electronic drives simply cannot go. This book makes reference to such traction drive applications as:

A 10,000-hp fixed-ratio helicopter transmission which is competitive with conventional technology geared transmissions.
A transmission for a 10-kW organic Rankine turbine which is part of a solar energy conversion system. The turbine speed must vary with operating conditions, but the alternator must operate at fixed output speed.
Small vehicle transmissions as are found in garden tractors and golf carts.
Machine tools driven by hydraulic motors.

There are many more applications in which traction drives that couple electric motors to loads have been selected in competition with other forms of drives. In these cases the traction drives were simply the best choice for the application. Electrically driven drives represent the majority of current traction drive applications.

The successful application usually involves the convergence of disciplines which require a good understanding of the capabilities of the prime mover, the requirements of the load, the characteristics of the traction drive, and the economics involved. This book is an excellent source for all of this information.

This is primarily a book for practical engineers who will be selecting, applying, and using these drives. The information is presented so as to be available at whatever depth the reader requires. There is enough reference material between these covers to allow practical drive selection and application for a wide range of uses. There may not be enough information to allow readers to *design* their own traction drives. Those who would follow this path may use this book as a starting point for appreciation as to what has already been done and for direction as to where to dig further.

Much of the ongoing work in the traction drive field comes from the metric countries. Chapter 4 will assist the reader in translating metric information to the customary system of measure in use in the United States. Chapter 8 contains information on standard metric motors and their installation interface dimensions, together with comparable information and torque characteristics of the standard NEMA motors.

Those who care to delve will find useful information on the basic theory of traction (Chapter 5) and on the interrelationship of the dominant variables that establish contact pad performance (Chapter 6). Chapter 7 develops the stress/life relationships of modern theory.

The cold realities of the selection process involve the mathematics of interest, taxes, and depreciation. These are discussed in Chapter 11, which presents interest and annuity tables which go high enough (25 to 28%) to be useful in a modern economy.

The workings of all the traction drives now available are described in Chapter 3. Appendix A is a listing of catalogs and addresses for the major traction drive suppliers worldwide. Appendix B, "Bibliography," is provided as a guide for further reading and research.

Chapter 12 is the heart of the book. It demonstrates, by the use of many real case histories, how the application process proceeds. It encourages the engineer to work interactively to change the problem so that it permits the best solution. This process, called systems engineering, is the way to forge a synergistic relationship between the prime mover, the drive, and the load which your competitors will find hard to shake.

2
HISTORY

2.1 INTRODUCTION

The traction drive got its start in the automotive industry to help solve the problem of transferring the power of the gasoline engine to the rear wheels. Development continued in an effort to have the engine run at its most efficient point under all road conditions.

The history of the drive and its hoped-for marriage with the internal combustion engine has been fraught with the continuing questions of cost of production and of durability. Success seemed very close three or four times, yet traction lost out to more conventional means.

Out of this continuing saga, a new field for the traction drive was cultivated. It is this industrial use that has enjoyed continued success and acceptance since the 1930s.

2.2 AUTOMOTIVE HISTORY

In less than a century, the gasoline-powered automobile has become a most dominant element in world society. It has literally created the auto and petroleum industries as major employers. It became the world's primary means of transportation, and at the same time one of its greatest sources of air pollution and its largest user of petroleum distillate [1].

The internal combustion engine has been the preferred source of propulsion over steam and electricity in spite of one basic and persistent limitation, its lack of power at low engine speeds. It must be used with a variable-ratio transmission of some form to permit the engine to run at a reasonable speed whatever the car speed.

The major advantage of the steam and electric drives was, and is, that they developed maximum torque at stall when needed to accelerate the car from standstill. The factors in favor of the gasoline engine were its portable power source, instant startup (electric also had this advantage), low cost, high power-to-weight ratio, long range without refueling, and high car speeds (steam also had high speeds, as evidenced by a Stanley Steamer holding the world speed of 197 miles per hour in 1907).

In early cars, friction transmissions delivered the variable speed ratio needed, but they lacked durability and the ease of automatic control. In the early 1900s, selective gear and planetary transmissions provided two to four selectable ratios, but required skilled drivers. The hydramatic transmission, which became available in the late 1930s, provided automatic selection, but at some loss in efficiency.

Today, through increased technology, hardened steel rolling surfaces in lieu of gears produce durable, continuously variable traction drives. These modern drives have automatic control, high performance because of their use of a synthetic traction fluid, and competitive cost. Whereas the beginnings of these drives was the search for a better automobile transmission, the results of these efforts was the development of the modern industrial traction drive, which has gained widespread acceptance in many industries.

The next several pages will trace the development of the automobile transmission from the first friction drives, with their power and durability limitations, to the near successes of the rolling contact traction drive in 1933 to the present advanced designs using modern metallurgy and lubrication.

2.2.1 Early Developments

The early gasoline-driven automobile had the problems of slow speed and limited power at low engine speeds when starting or hill climbing. This required a speed-changing transmission and a clutch to disengage the engine from the drive wheels during startup, idling, and gear changing. This was the only way to match the engine output speed and torque with that required for the road.

Various ratio-changing transmissions were used. These took the form of friction drives, selective sliding gearboxes and progressive sliding gearboxes with clutches, planetary gears with brake bands, and belt systems with loose pulleys.

The friction disc transmission had a large future. The friction transmission consisted of two elements, the driving friction disc and the driven friction disc. The simplest form had the driven disc set on a shaft at right angles to the driving disc. It was rotated by friction contact between its edge and the face of the driver. When the edge of the driven disc was driven on a circle nearest the periphery

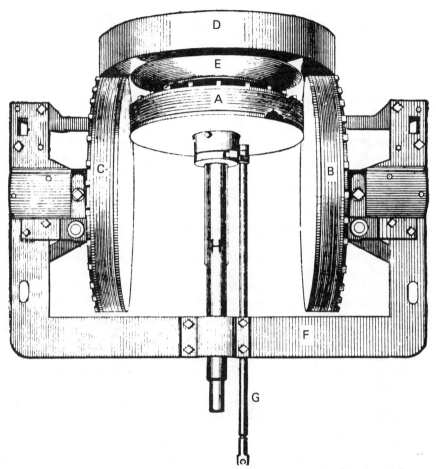

FIG. 2.1 Schematic of a typical early 1900s automobile friction drive transmission. (From Ref. 2; courtesy of the Bobbs-Merrill Company, Indianapolis, Ind.)

of the driver, its speed was greatest. As it was slid along its shaft toward the center of the driver, its speed was constantly decreased. At the center of the driving disc it ceased to rotate. When slid beyond the center of the driver, its motion was reversed.

Figure 2.1 shows a type of friction transmission. A is the driven disc on the transmission shaft; B and C are friction idlers driven from the driving disc D; E is the clutch; F is the transmission frame; G is the lever for changing the speeds by shifting disc A along shaft H.

The 1891 Lambert, a three-wheeler with a two-cylinder engine, used a friction transmission which served as a combination clutch and variable-ratio drive. The drive consisted of a metal friction face on the back side of the flywheel and a friction wheel on a splined cross shaft. The wheel had a friction board surface on the rim that was pushed against the face of the flywheel by a foot pedal and held in contact by a ratchet. When the ratchet was released, the friction wheel could be moved across the flywheel by a hand lever to provide the desired speed change and reverse.

In 1892, the Duryeas operated their first car with a friction transmission, but later changed to a two-speed constant mesh transmission. The 1896 Reeves used variable-pitch pulleys for speed changing. Ford's Quadricycle used a belt and idler drive engaged by moving the idler against the slack side of the belt. R. E. Olds offered his cars with a two-speed planetary gear transmission with brake bands. Winton's 1897 auto used a two-speed constant mesh gear transmission with a clutch. The friction transmission with its infinitely variable ratio provided a solution to match the engine's speed and torque characteristics to the desired driving conditions.

In 1907, *Horseless Age* (Feb. 20, p. 267) announced: "Frictional transmissions are simple in construction, not likely to be mishandled, of a generally rugged character, and not costly to manufacture." Unfortunately, when the friction transmission cars of those days were operated at low speed, efficiency was poor, and wear of the mating parts was high. Even so, the friction transmission remained in use until World War I before it was replaced by the standard transmission and clutch.

Greater engine power and space limitations imposed by the diameter of the friction disc, and problems of slip and wear all placed limitations on the friction drive and were responsible for its discontinuance. A solution to wear on the friction surfaces was never found, especially in the face of increased contact pressure required by the increased engine torque.

Many early automobile models had short lives, with few companies being in business longer than two or three years. Among the longer-lived manufacturers, the following used friction drive transmissions [3,4].

The Lambert was manufactured from 1891 to 1917 by the Buckeye Manufacturing Co., Anderson, Indiana (Fig. 2.2). The friction disc was 18 in. in diameter.

The Lewis was manufactured from 1898 to 1901 by the Lewis Cycle Co., Brooklyn, New York. The car had a single-cylinder horizontal engine with the friction disc surface made of compressed paper.

The Tourist was manufactured from 1902 to 1910 by the Auto Vehicle Co., Los Angeles, California. The two-cylinder water-cooled model used a friction transmission.

Automotive History

FIG. 2.2 Advertisement for the 1907 Lambert, which featured a friction drive transmission. (Courtesy of Bonanza Books, New York.)

The Cannon was manufactured from 1902 to 1906 by the Burtt Manufacturing Co., Kalamazoo, Michigan. The two-cylinder tonneau used a friction drive.

The Marble-Swift was manufactured from 1903 to 1905 by the Marble-Swift Automobile Co., Chicago, Illinois. It was a two-seat two-cylinder front-engine car with a friction drive.

The Cartercar was manufactured from 1906 to 1916 by The Motor Car Co., Pontiac, Michigan. In 1909, the company became part of General Motors. All the early models used a friction transmission (Fig. 2.3).

The Sears was built from 1906 to 1911 by the Sears Motor Car Works, Chicago, Illinois (Fig. 2.4). It was built for Sears Roebuck catalog sales and some 3500 were sold by mail order. It was a high wheeler with a two-cylinder horizontal air-cooled engine. The lever in front of the steering column moved a fiber-faced wheel across the drive disc. A foot pedal held the wheel against the disc.

The Simplicity was built by the Evansville Automobile Co., Evansville, Indiana, from 1906 to 1911. It was a four-cylinder water-cooled roadster and touring car with a friction drive and double chain drive.

FIG. 2.3 Early 1900s Cartercar advertisement, featuring its friction drive. (Courtesy of Bonanza Books, New York.)

FIG. 2.4 1910 Sears Model K, which featured a selective friction drive transmission. (Courtesy of Digest Books Inc., Northfield, Ill.)

Automotive History

The Holmes was built from 1906 to 1908 by the Holmes Motor Vehicle Co., East Boston, Massachusetts. They were five-seater touring cars with two- and four-cylinder engines and a Reeves friction transmission.

The Victor was produced from 1907 to 1911 by the Victor Automobile Manufacturing Co., St. Louis, Missouri. It was a high wheeler with a single-cylinder water-cooled engine with a friction drive.

The International manufactured by International Harvester, Akron, Ohio, from 1907 to 1911 was a two-seat surrey or a light delivery truck with a two-cylinder engine with a friction transmission.

The Earl was built from 1907 to 1908 by the Earl Motor Car Co., Kenosha, Wisconsin. It was a light roadster with a friction drive.

The Stanley was built from 1906 to 1908 by the Stanley Automobile Co., Moreland, Indiana. It was a five-passenger touring car with a two-cylinder water-cooled engine with a friction drive.

The Burns, built from 1908 to 1911 by Burns Brothers, Havre de Grace, Maryland, was a high wheeler with a two-cylinder air-cooled engine, friction transmission, and double chain drive.

The Gearless was built from 1907 to 1909 by the Gearless Motor Car Co., Rochester, New York. They were high-powered cars with a friction transmission. The transmission of the large Gearless cars was of the planetary friction type (Fig. 2.5). There were two forward speeds and one reverse, without the use of any gears, the high speed being direct, in which the speed-change elements revolve together as a unit with no internal friction or rolling contact. The entire speed-change unit revolved together as a flywheel. It consisted of six large special fiber rolls of conical shape revolving on and in an exterior and interior cone. These two cones co-acted with a sliding, double-faced, solid jaw clutch, which was moved to the extreme forward position to give the low speed forward, and to the extreme rearward position to give the reverse. The internal cone was constantly pressed toward the external by means of a spring so as to always ensure "bite" enough to make the six cone rollers revolve without slipping in the low-speed and reverse drives.

The Gearless transmission had the advantage of no change gear friction at the high speed or direct drive and rolling friction engagement in the low speed and reverse.

The Pontiac, manufactured from 1907 to 1909 by the Pontiac Spring and Wagon Works, Pontiac, Michigan, was a two-seater high wheeler with a two-cylinder water-cooled engine, friction transmission, and double chain drive. It was taken over by Cartercar in 1909.

The Metz, built from 1908 to 1922 by the Metz Co., Waltham, Massachusetts, had a friction transmission until 1919 (Fig. 2.6). Originally, it was a two-cylinder light roadster which was sold for home assembly until 1912, when it was replaced by an assembled four-cylinder car.

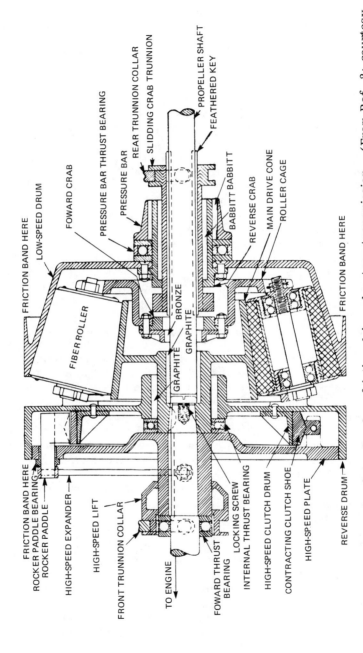

FIG. 2.5 Schematic of the Gearless friction drive planetary transmission. (From Ref. 2; courtesy of the Bobbs-Merrill Company, Indianapolis, Ind.)

FIG. 2.6 Advertisement for the 1914 Metz, "The Gearless Car." (Courtesy of Bonanza Books, New York.)

The Ames, built from 1910 to 1917 by the Ames Motor Car Co., Owensboro, Kentucky, was available as a roadster and a five-seater with a four-cylinder engine and friction transmission.

A score of other makes and models with friction drive appeared briefly and passed on, including a number of lightweight cycle cars with friction drive that were popular just before World War I. Only one new friction drive car was introduced after World War I ended. It was the Kelsey, built from 1921 to 1924 by the Kelsey Motor Co., Belleville, New Jersey.

The Pittsburgh electric truck manufactured by the Pittsburgh Motor Vehicle Co., Pittsburgh, Pennsylvania, was one of several trucks to use a friction transmission. Power was transmitted (Fig. 2.7) by leather fiber cones coming into contact with a bevel cast iron wheel. The front cone gave the reverse motion to the truck, the intermediate cone the high speed, and the rear cone the slow speed.

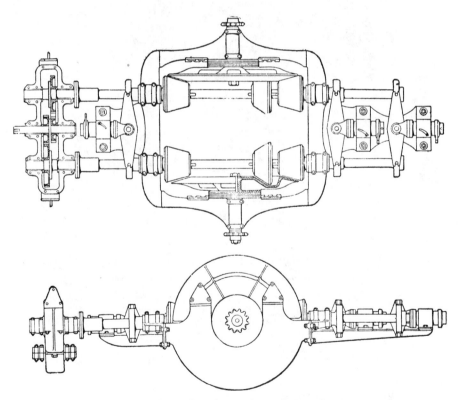

FIG. 2.7 Schematic of the friction drive of the Pittsburgh truck. (From Ref. 2; courtesy of the Bobbs-Merrill Company, Indianapolis, Ind.

The advent of the synchromesh transmission in 1928 removed the problem of clashing gears during shifting of the manual transmission and brought it to its final stage of development. The search was still on for a means of automatically changing gears to match the engine's power with the road requirements.

Richard Erban in Austria developed a continuously variable form of traction drive for a 40-hp auto in 1924 and test drove it through Europe. He applied for a U.S. patent in 1922 and received it in 1926. His patent included six different methods for changing ratio, all of which used a torque-responsive system to increase the rolling contact pressure with increasing load and thus reduce the contact stresses during light loads. Erban's traction drive transferred torque by rolling contact through a thin film of lubricant trapped in the contact area under high fluid pressure. The rolling contact was similar to that of the friction drive; however, the torque transfer was not by friction, but by shear in the thin film of lubricant at the point of rolling contact. This phenomenon later became known as elastohydrodynamic lubrication. Peugeot acquired the license for automobile use in 1927 for use in a 25-hp automobile.

In 1925, Frank Hayes applied for a patent covering a dual toroidal drive that required no external thrust bearing to induce the rolling contact pressure. The rollers were steered to change the ratio, and the output speed ranged down to zero and through zero to reverse. He received his patent in 1929. Hayes installed his drive in a bus and demonstrated it in 1929. Austin of England was interested enough in the transmission to offer it with a hydraulic automatic control as an option. The system was dropped after one year, since it was relatively expensive, and it exhibited durability problems and excessive traction fluid contamination.

In 1925, General Motors was studying nine types of transmissions, including four continuously variable drives (i.e., variable throw, inertia, hydraulic, and electric), but none of these held the promise of meeting the cost, noise, size, or efficiency criteria established for the study.

In 1928, Alfred Sloan, president of General Motors, asked Charles Kettering, head of G.M. Research, to develop an automatic transmission suitable for mass production (inexpensive and easily manufactured) that would improve engine performance and car drivability when operated by any driver. Various traction drive configurations were examined and the dual toroidal drive was selected.

This drive promised adequate ratio range, control response time equal to engine response, adaptability to automatic control, good efficiency, acceptable size and weight, durability in the 50,000-mile range, and competitive cost. The design chosen had as a major feature the precise control of the roller position and reduction of spin at the roller contact, which reduced losses due to friction. A centrifugal clutch was employed for starting and stopping. The overall range was 8:1, 3.2:1 reduction to 1:2.5 for overdrive.

General Motors engineers set out to derive a method of keeping the rollers in position to share the load, obtain rapid response to ratio change commands, increase the durability of the rolling contact surfaces subjected to high Hertz stresses, overcome limitations on the torque capacity imposed by the modest traction coefficient of the lubricant, and define a means to automatically control the system to match engine speed with the road conditions.

The completed design had an infinitely variable ratio control which was fully automatic with throttle control. It weighed 250 lb. It had an operating efficiency of 87 to 91%, and it enabled the vehicle to exhibit a 20% increase in fuel economy compared to the existing synchromesh transmissions.

At the conclusion of tests in 1933, Buick expressed an interest in the dual toroidal transmission and assumed the design responsibility for the production unit, which was completed in 1935. Despite its greater cost compared to a standard transmission and some unsolved durability problems, it was offered as an option. Its introduction was canceled in favor of a four-speed planetary transmission with a fluid coupling. Cadillac entered a 160-hp toroidal drive in the 1934 Pikes Peak run, but the drive experienced excessive wear.

General Motor's New Departure Division undertook further development of the line for nonautomotive use in 1935 and named the drive Transitorq. The drive was powered by a constant-speed electric motor. It featured manual ratio control over a ratio range of 6:1, and the capability to change this ratio at rest or when in motion. The drive would automatically return to low speed when it stopped. Its selling points were continuously variable ratio, freedom from wear, and the absence of gear noise and vibration. Durability of the units was good, but the costs were higher than other mechanical adjustable-speed drives, and production ceased in 1937 after more than 2500 units were sold to machine tool builders.

Cost and durability were interrelated throughout the development history. Output torque was limited by the fatigue resistance of the hardened steel rollers and the traction coefficient (tangential force/normal force) of the available lubricants. This led to modest ratings for any given size.

During 1937, Perbury in England developed a dual toroidal drive with rollers steered for ratio change to balance engine speed with load. They also developed the elastohydrodynamic concept to account for the transfer of torque through the lubricant film.

In the 1950s, while president of Kraus Engineering Company, Charles E. Kraus began experimenting with continuously variable traction drives for automobile use as a hobby and developed a 90-hp

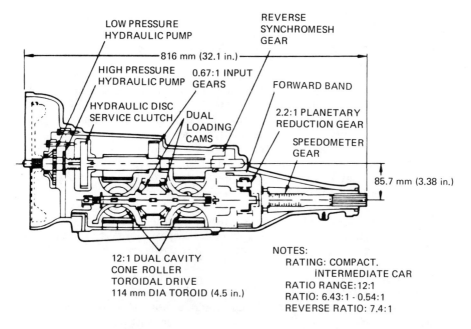

FIG. 2.8 Schematic of the Traction Propulsion Inc. traction drive continuously variable transmission. (Courtesy of Aerospace Corp., El Segundo, Calif.)

toroidal traction drive. He received a patent in 1958 on a continuously variable toroidal drive, with the ratio responding to the torque load independent of speed. He installed this unit in a Rambler, accumulating 10,000 miles of testing by 1959. Curtiss-Wright continued development of the transmission with support from American Motors.

In 1973, under the sponsorship of Tracor in Austin, Texas, a single toroidal element with cone rollers and hydrostatic bearing for necessary roller contact pressure was installed in a Ford Pinto demonstrator. Since 1976, the development has been continued by Traction Propulsion, Inc. of Austin, Texas (Fig. 2.8).

In the late 1950s, Avco and Pratt & Whitney were also active in the development of a drive for automotive use and Avco continued development for constant-speed aircraft alternators.

In 1953, a review of G.M. research data from the 1930s on the traction drive [5] showed that by updating the earlier design with improved materials and lubricants, the durability could be increased to 100,000 miles and fuel consumption could be cut by 50% of the then current usage for normal driving. This review also showed the traction drive transmission to be optimum for a vehicle designed for smaller size, lighter weight, and less drag.

In an analysis of this material [1], the following problem areas were identified.

1. Lack of consistency between the predicted performance and the 1937 test data.
2. Any design would have to recognize that the 100,000-mile durability required necessitated an endurance life greater than that available from steels previously used.
3. Improvements were needed in the hydraulic roller control system, and a more rigid case would be needed, to allow multiple rollers to share the load equally.
4. A better understanding was required of the traction phenomenon in which torque transfer in rolling contact depended on elastohydrodynamic shear resistance of the layer of lubricant trapped in the contact area under pressure.

A transmission was designed using new bearing steel and a more rigid transmission case. This drive could be made for a competitive cost, but no suitable traction fluid could be found for it. Test transmissions were built from 1957 to 1959 and demonstrated the better efficiency and fuel economy and acceleration promised. Drivability turned out to be harsh. Further work was undertaken in 1963 to improve the drivability with a full-time fluid coupling added to the toroidal transmission and to make the unit more cost effective without the loss of efficiency.

During the 1960s, most automotive emphasis was on high performance. In 1970, the Environmental Protection Agency mandated emission standards which placed restrictions on the internal combustion engine that severely limited the automobile's high-compression engine.

G.M. Research started a new study on the gas turbine engine as a way to reduce pollutants. The two-shaft gas turbine with a standard transmission seemed most compatible with the conventional automobile power train. The single-shaft turbine, which would have been a simpler and less costly engine, needed a continuously variable transmission to deliver the necessary performance and fuel economy. In 1971, G.M. Research built and tested both systems. The program was eventually canceled due to unsolved durability, performance, and cost problems.

With fuel-efficient vehicles being mandated by legislation, a new era of American automobiles has begun to appear. Reduced weight and engine power has made the variable-ratio transmission ripe for introduction. The space under the hood is inadequate for a transverse engine with a traction drive. A new design approach will be needed to combine the engine and transmission into one unit to reduce weight, number of parts, cost, and power-consuming supportive systems.

2.2.2 Fluid Development

One of the difficulties encountered with all the early work done on the traction drives was the limited ability of the traction fluid to withstand the tremendous pressures experienced when the thin layer of lubricant was squeezed between the rolling surfaces and limited the traction coefficient. Most of the effort reported earlier was focused on metal qualities and manufacturing potential (i.e., durability and new material availability), but no improvement was made to the traction fluid.

The fluids industry was approached by G.M. Research to develop a fluid with a higher coefficient, and Monsanto picked up the challenge. By 1967, Monsanto had developed a new synthetic traction lubricant with a 30% higher traction coefficient. It was expensive, but it worked. It was supplied to G.M., which was developing a passenger bus for the U.S. Department of Transportation using a gas turbine power plant with a dual toroidal transmission and fluid clutch. The new fluid and transmission were successful, but due to difficulties with the turbine, G.M. reverted to the conventional diesel engine and its commercial transmission.

By 1969, Monsanto had completed research on 30 lubricants, including some new synthetics, and found that one, cycloaliphatic hydrocarbon, exhibited a 50% increase in the traction coefficient over that of petroleum oil. In addition to the higher traction coefficient in the elastohydrodynamic range, the fluid was unaffected by continuous operation at 200°F. Its glass transition pressure was in the range of 100,000 psi. It offered excellent lubrication for the sliding and rolling surfaces. Santotrac® was tested by Timken, New Departure, and SKF and showed increased fatigue life of the rolling surfaces under Hertz stresses encountered by roller bearings. This higher coefficient offered an answer to the automotive traction drive by allowing higher torques to be delivered in a smaller package at less price.

2.2.3 Modern Developments

In addition to the continued efforts of Charles E. Kraus to develop a successful automotive application for the traction drive, at least 10 others have, for the past decade and into the 1980s, been seriously pursuing the idea of developing a traction or friction drive for automotive-related use. These include, but are not limited to, the following.

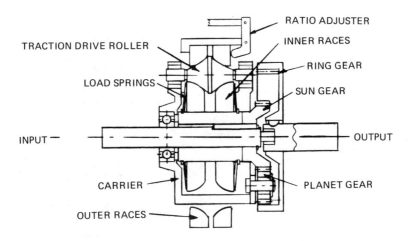

FIG. 2.9 Schematic of the Fafnir Bearing variable-ratio planetary tractor transmission. (Courtesy of Fafnir Bearing, Division of Textron, New Britain, Conn.)

Fafnir Bearing is using a planetary ring and roller design, originally developed by Walter Chery and the Arthur D. Little, Inc., for use in their agricultural equipment transmission (Fig. 2.9). The drive consists of a double conical roller that drives a planetary gear cage. The traction roller rides between two two-part races: an inner race is splined to the drive shaft and a stationary outer race. The drive is hydraulically controlled by spreading or contracting the outer race halves and thereby changing the traction roller's relative speed.

Yves Kemper, president of Vadetec Corporation, has developed a traction drive (Fig. 2.10) with fewer parts than the conventional automatic transmission. The drive is used with an energy storage system, called Inertial Drive Line, and a microprocessor. The new drive is a sophisticated version of the ring and cone in which the main shaft nutates or wobbles, enabling it to handle high horsepower. The drive's input shaft is attached to a rotating cylinder, and set into the cylinder is a freely turning double cone roller. As the input shaft turns, the cone roller is forced to roll along the adjustable traction rings. As the cone rotates, a planetary gear on the output end drives the output shaft. The speed is varied by shifting the adjustable traction rings back and forth on the cone over its varying radii.

British Leyland, now BL Technology Ltd., is working with a modified Perbury roller toroidal drive with a planetary gear box mated to it (Fig. 2.11). The rollers link one rotational radius on the input

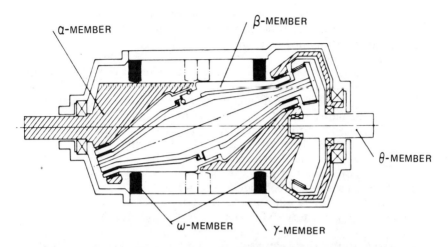

FIG. 2.10 Schematic of the Vadetec continuously variable transmission. (Courtesy of Vadetec Corporation, Troy, Mich.)

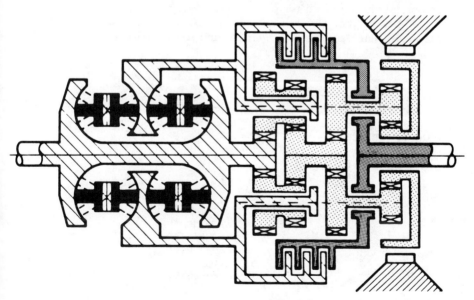

FIG. 2.11 Schematic of the British-Leyland/Perbury continuously variable transmission. (Courtesy of Aerospace Corp., El Segundo, Calif.)

side of the toroidal cavity with a second rotational radius on the output. By tilting the rollers, the relationship between the radii is altered giving the varying ratio.

Borg-Warner Corp., in conjunction with Van Doorne Transmissie, is perfecting a friction drive transmission which uses hydraulically controlled, variable-diameter pulleys and a steel V-belt arrangement (Fig. 2.12). The steel belt is constructed of steel blocks held in a loop by a laminated steel band which is pushed rather than pulled by the drive pulley, using the metal-compression principle.

Ingersoll Engineers is working on a prototype design.

Battelle Columbus Laboratories is working on a twin steel push-type system on the same principle as Borg-Warner, with the pulleys being hydraulically controlled (Fig. 2.13).

Garrett Corp. is experimenting with an electric flywheel hybrid vehicle using a full toroidal cavity drive with steerable rollers (Fig. 2.14).

Kumm Industries is proposing a power-split flat belt configuration with the ratio being changed with hydraulically controlled, fixed-width, adjustable-shaft pulleys (Fig. 2.15).

The Bales-McCoin Inc. design uses the cone roller principle and operates on the premise that as speed increases, the traction coefficient drops and the required torque decreases (Fig. 2.16).

Gates Rubber Co. has developed a compact ring-cone drive with a traction planetary for light vehicle and industrial use.

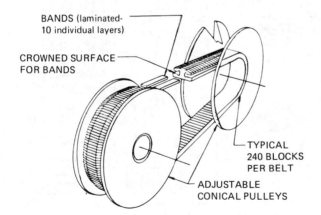

FIG. 2.12 Schematic of the Borg-Warner/Van Doorne steel V-belt. (Courtesy of Aerospace Corp., El Segundo, Calif.)

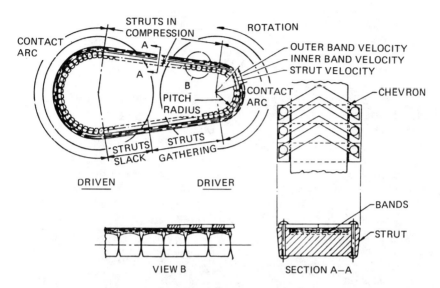

FIG. 2.13 Schematic of the Battelle Columbus Laboratories steel V-belt drive. (Courtesy of Aerospace Corp., El Segundo, Calif.)

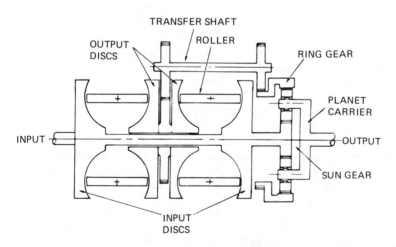

FIG. 2.14 Schematic of the Garrett Corp./AiResearch Div. traction drive continuously variable transmission. (Courtesy of Aerospace Corp., El Segundo, Calif.)

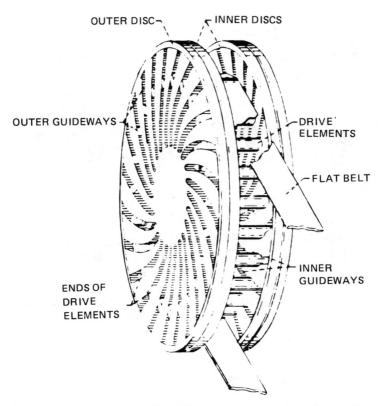

FIG. 2.15 Schematic of the Kumm Industries continuously variable transmission. (Courtesy of Aerospace Corp., El Segundo, Calif.)

2.3 MODERN INDUSTRIAL DEVELOPMENTS

The early friction drives consisted of leather-faced rollers in contact with metal discs. One of the earliest patented American drives was invented by Charles Hunt in 1877 (Fig. 2.17). This was a toroidal-type unit that was used with belt-driven countershafts.

Much of the early traction drive work that was initiated for automotive use found its way into industrial applications such as the unit developed by General Motors, which became the New Departure Transitorq. Jacob Arter began marketing a ring-cone traction drive for industrial applications in Switzerland in 1926. AEG-Regulies Getriebe of Germany marketed a 5.5-hp toroidal drive in 1929. This drive used a lubricant similar to SAE 10 which had a traction coefficient of 0.05 and an efficiency of 93%.

Modern Industrial Developments

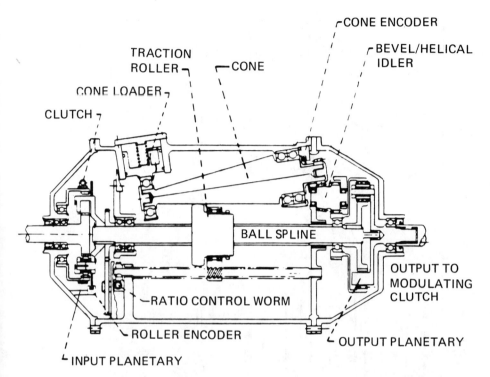

FIG. 2.16 Schematic of the Bales-McCoin Inc. traction drive continuously variable transmission. (Courtesy of Aerospace Corp., El Segundo, Calif.)

Lou Graham started his original work on traction drives while employed as an engineer for Relay Motors and then Falk, to develop a drive as a replacement for the Model T transmission, but the Model T Ford ceased production in early 1927. In 1935 he developed a continuously variable ring roller traction drive that included planetary gearing to give forward and reverse speeds. He received a patent in 1936. In 1941, Briggs & Stratton took over the manufacture and the improvement of the Graham drive. The Graham drive found wartime applications in gun controls and camera strip-film drives. After the war, industrial applications were pursued, with further refinements being made. The Graham ring roller with planetary is still in production.

The modern industrial traction drive got its real start in 1945. Jean E. Kopp, a young engineer, was working for an aircraft company in Berne, Switzerland. He developed a drive for variable-pitch propellers which used a hard steel ball as the main transmission component.

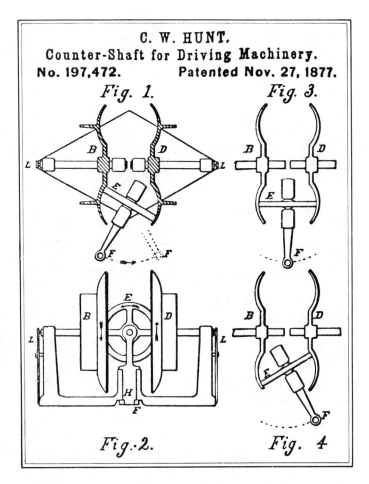

FIG. 2.17 1877 patent drawing of Charles Hunt's toroidal drive.

His employer told him that his drive would have limited demand since the production of propeller-driven planes would decrease due to the advent of the jet-propelled aircraft.

Kopp continued to work on his idea and saw that it could be used in a variable-speed power transmission. He quit his job with the aircraft company to perfect the design, have prototypes manufactured, and to patent his drive.

In 1949, Allspeeds in England was licensed to produce the ball variator, and in 1954, Eaton was licensed to manufacture it in the United States for industrial use. Kopp continued to refine the basic ball variator by changing the manner in which the balls were steered.

Kopp developed a roller variator which used tapered rollers rather than balls. This drive attained an efficiency of 83 to 94% depending on the speed ratio. Koppers Company was licensed as the American manufacturer.

Charles E. Kraus started work on his traction drives in the early 1950s. His developments were directed toward the automotive market, but his toroidal drive and its offshoot spool drive also had some industrial applications.

The Graham, Kopp, and the Kraus drives are discussed further in Chapter 3. In addition, the drives developed by Arter, Beier, Contraves, Floyd, Hans Heynau, Shimpo, Sumitomo, Unicum, Vadetec, and others are examined in detail in Chapter 3.

A fixed-ratio traction drive to be discussed later that was invented by Algirdis Nasvytis and further refined by the National Aeronautics and Space Administration (NASA) was adapted for an application in a helicopter transmission. This drive is a fixed-ratio planetary capable of large speed increase or reduction ratios operating at output speeds of 480,000 rpm. These drives are extremely compact and lightweight and still able to transmit high torques. The solution is in stepping the planetary rollers such that they carry the power through two different diameters. Input may be applied to either the central sun roller for speed reduction or to the output ring rollers for speed increase.

REFERENCES

1. R. W. Carson, *Continuously Variable Transmissions for Automobiles*, unpublished manuscript prepared for Monsanto, St. Louis, 1977.
2. J. E. Homans, *Self Propelled Vehicles, a Practical Treatise on the Theory, Construction, Operation, Care, and Management of All Forms of Automobiles*, Theodore Audel & Co., New York, 1910.
3. *The American Car Since 1775*, An Automobile Quarterly Library Series Book, Automobile Quarterly Inc., New York, 1971.
4. G. N. Georgano, *Complete Encyclopedia of Motor Cars—1885 to the Present*, Rev. Ed., E. P. Dutton & Company, Inc., New York, 1973.
5. D. F. Caris and R. A. Richardson, "Engine - Transmission Relationships for High Efficiency," *SAE Trans.*, vol. 61, 1953, p. 83.

3
GENERIC TYPES

3.1 TYPES OF TRACTION DRIVES

Traction drives can be categorized by the generic types shown in Fig. 3.1.

3.1.1 Kopp Variator

Ball Variator

The first Kopp ball variator, available from Jean E. Kopp Variators, or from Winsmith, Allspeeds, Cleveland Gear, and other licensees, was introduced in 1947 by Jean Kopp of Mürten, Switzerland, and within a few years was being successfully manufactured under license in many countries. The Kopp design (Fig. 3.2) is said to be the most widely used type of adjustable-speed traction drive. More than 1 million units have been produced. The original ball variator is still in production.

Operating features

1. The output speed of the Kopp ball variator is adjustable from one-third to three times the input speed for a 9:1 overall speed range. An updated ball variator has a speed range of 8:1 for constant horsepower and up to 12:1 for constant torque applications.
2. Except for minor efficiency variations, the output horsepower of the variator is constant throughout the speed range. Constant-torque application requirements can be satisfied by selecting a unit capable of providing the required torque at the highest required output speed.

Types of Traction Drives

3. A built-in device adjusts the pressure between drive balls and cones in proportion to load.
4. Under extreme no-load-to-full-load fluctuation, maximum output speed change is under 4%. Under uniform loading, output speed variation is less than 0.1%.
5. On restart, it returns to the last set ratio and provides accuracy of repeatability within 0.1% with manual vernier control.
6. The drive's efficiency is 90% at 1:1 ratio and drops to 82% at the maximum ratio.
7. The ball variator is available in sizes from fractional horsepower to 16 hp.
8. The speed ratio cannot be changed when the unit is at rest. The variator must be rotating to adjust the ratio.
9. Either shaft can be used as the input shaft.
10. The ball variators can be operated either clockwise or counterclockwise.
11. Foot-mount and flange-mount styles are available.
12. Gear reducers are available.
13. The output speed of the basic unit is 600 to 5400 rpm at 1800-rpm input.
14. The original version of the drive must be horizontally mounted only, but the updated model can be made for vertical mounting when specified.
15. A limitation of the Kopp ball variators is that they must be protected from shock loads.

Principles of operation. The sectional illustration (Fig. 3.2) is of a typical ball variator. The outer casing consists of a body and two end covers. The coaxial input and output drive shafts rotate in the same direction. Power is transmitted from the input drive shaft (C) through a disc (A) splined onto the shaft by means of a pressure device (B). Power passes to the input drive cone (D) and then to a series of drive balls (E) mounted on ball spindles (K) which incorporate two sets of needle roller bearings (J). The ball spindles are located at each end in radial slots in the end covers, passing through a series of cam-shaped slots in an iris plate (F) which can be rotated by the control worm (G). This movement tilts the axes of the drive balls and produces the speed variation.

From the drive balls (E) power is transmitted through the output drive cone (L), by a second pressure device (N), to the disc (P), which is splined to the output drive shaft (M). The drive balls (E) are held in position and in contact with the drive cones (D) and (L) by a free-floating retaining ring (H).

Output speed is determined by relative lengths of the contact paths on input and output sides of the balls. By tilting both axles

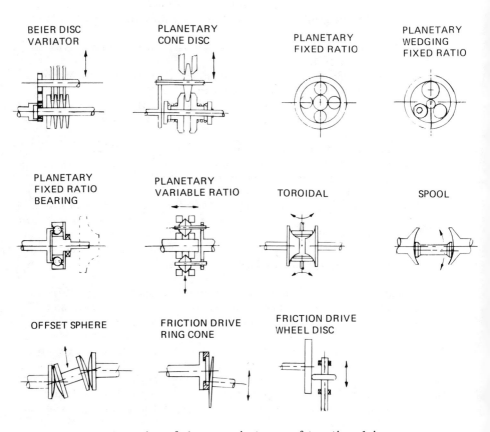

FIG. 3.1 Schematics of the generic types of traction drives.

Types of Traction Drives

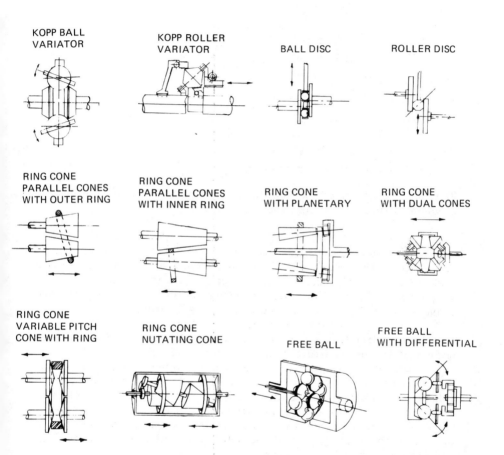

FIG. 3.1 (Continued)

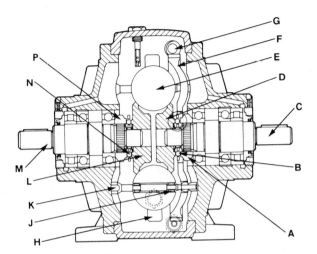

FIG. 3.2 Schematic of the Kopp ball variator. (Courtesy of Winsmith, Division of UMC Industries, Inc., Springville, N.Y.)

and balls, relative lengths of the two contact paths are varied. In Fig. 3.3A, both contact paths are equal in length, thus giving a speed radio of 1:1. In Fig. 3.3B, ball contact path on the input side is longer than on the output side, resulting in a decrease in output speed; conversely, in Fig. 3.3C, an increase in speed results.

A torque-responsive mechanism (Fig. 3.4) is incorporated into both input and output shafts to provide traction proportioned to the amount of torque transmitted through the drive.

When torque is applied to either the drive shaft or the drive cone, it is transmitted from one to the other through six or eight spherical rollers (S) secured in a retainer and normally centered in the shallow V notches (R) formed on the opposing faces of the splined ring and the drive disc. In transmitting this torque, the spherical rollers tend to ride up the sloping sides of the notches and force the splined ring and the drive disc farther apart.

The thrust bearings in the end plates of the unit are adjusted to remove all end play; any actual widening of the space between the splined ring and the drive disc is limited by the stiffness of the whole structure. An axial pressure is established in the unit that is transmitted from one thrust bearing to the opposing one through the drive discs and balls. The axial pressure so created is proportional to the applied torque and is adequate to provide positive traction between the drive discs and the balls. This proportionality between torque and axial pressure is maintained even under severe overload and is independent of speed.

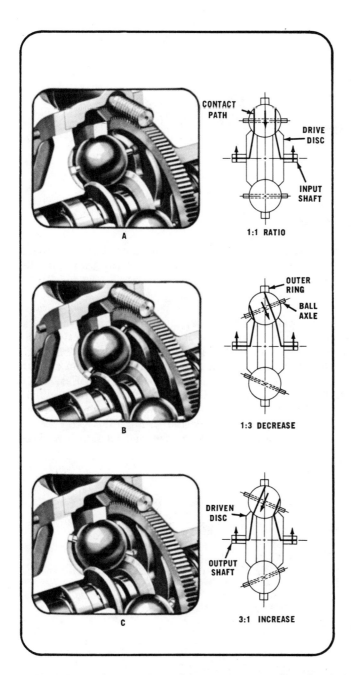

FIG. 3.3 Variator ball position and power flow for 1:1, 1:3, and 3:1 speed ratios. (Courtesy of Cleveland Gear, Cleveland, Ohio.)

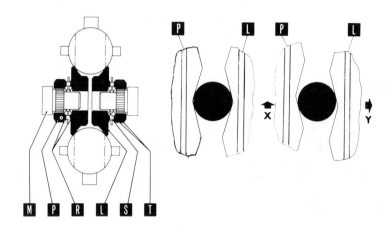

FIG. 3.4 Torque-loading cam mechanism for the Kopp ball variator. (Courtesy of Winsmith, Division of UMC Industries, Inc., Springville, N.Y.)

At the instant of starting the drive, a nominal axial pressure is required to ensure that the drive discs and balls are in tractive contact. This is supplied by a series of Belleville springs (T). The ball variator will not slip, but will stall an equally rated induction motor.

Figure 3.5 shows an exploded view of a complete drive assembly. Figure 3.6 shows the change in output speed with change in applied load as plotted for different ratio settings in the ball variator. The efficiency of a typical ball variator is plotted in Fig. 3.7. In this case, the curve is representative of a standard ball variator operated

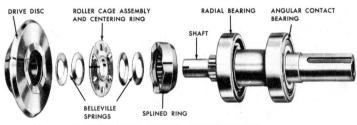

FIG. 3.5 Exploded view of the complete torque-loading assembly. (Courtesy of Cleveland Gear, Cleveland, Ohio.)

Types of Traction Drives

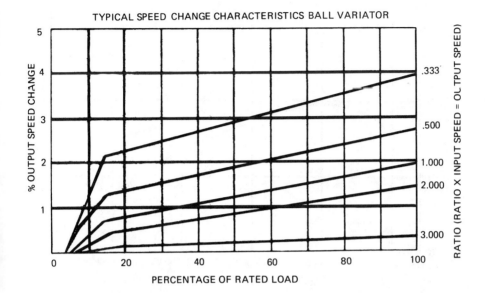

FIG. 3.6 Typical speed characteristics for the Kopp ball variator. (Courtesy of Cleveland Gear, Cleveland, Ohio.)

at 1200-rpm input speed, with output speeds varying from 400 to 3600 rpm. Maximum efficiency is attained when input and output speeds are equal.

Type M Variator

The principle of operation for the updated, type M, Kopp ball variator is similar to the original version. The main difference is the method by which the balls are steered.

In Fig. 3.8, disc (6) rotates freely on shafts (3) and (4) in housing (1) with end cover (2). A thrust disc is fixed to shafts (3) and (4). The front face of disc (5) is provided with ramps. Identical ramps are provided on disc (6) and form, with the balls arranged between the ramps, a device for producing automatic pressure on disc (6) when torque is applied to the shaft (Fig. 3.9). Balls are in friction contact with disc (6) and abut against the spherical segment (9) via bearing (8). Each ball is formed with an annular groove, which forces the balls to rotate around the axis of the bearing. The ball is mounted to slide axially, but does not rotate in the housing. An adjusting screw engages the adjusting ring (10), which is shifted axially by rotation of the screw.

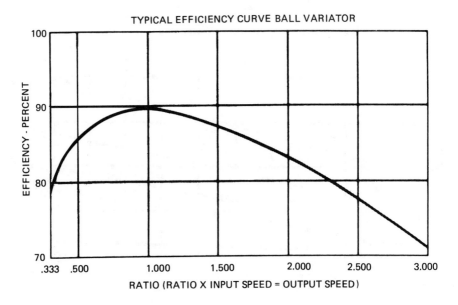

FIG. 3.7 Typical efficiency curve of the Kopp ball variator. (Courtesy of Cleveland Gear, Cleveland, Ohio.)

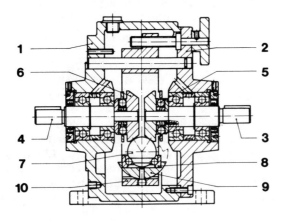

FIG. 3.8 Schematic of the Kopp type M ball variator. (Courtesy of Kopp Variatoren AG, Mürten, Switzerland.)

Types of Traction Drives

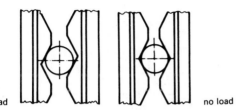

FIG. 3.9 Torque-sensing load mechanism of the type M Kopp ball variator. (Courtesy of Kopp Variatoren AG, Mürten, Switzerland.)

The spherical head of the control segment (9) engages the adjusting ring (10) so that, by axial shifting of this ring, the control segment changes its position, as shown in Fig. 3.10. In the middle position, the speed ratio between the two discs (5) is 1:1. Turning the control segment to the center of the ball produces a modification of the speed ratio between the two discs, which is due to the shorter distance between rotation axles of the ball and the point of contact of the ball with one of the discs.

Theoretically, an infinite speed ratio could be obtained when the axis of rotation of the ball coincides with the point of contact of the ball on one disc. In order not to overload the contact surfaces, the speed ratio is limited to 1:12 in standard application. The type M ball variators can be obtained in configurations that allow vertical and horizontal mounting.

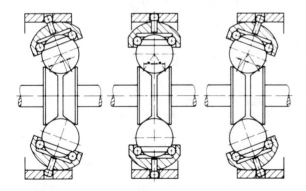

FIG. 3.10 Schematic of the typical ball positions for speed-change ratios. (Courtesy of Kopp Variatoren AG, Mürten, Switzerland.)

Roller Variator

In 1959, Jean Kopp began to produce a tapered roller variator. These units are able to transmit up to 100 hp and had an output range of 12:1 (7:1 reduction ratio and a 1.7:1 increase ratio).

In 1974, Koppers Company, Inc., was licensed to manufacture and sell the roller variator in the United States, Canada, and Mexico.

Operating features

1. With an input speed of 1800 rpm, the output speed is adjustable from 250 to 3000 rpm.
2. The high power capacity of the roller variator in sizes up to 100 hp is achieved by distributing the power transmission over a large number of rollers arranged in a circle around the input and output power elements in a manner similar to that of a planetary gear.
3. The speed ratio gives reduction ratios up to 7:1 and increaser ratios to 1:1.7.
4. The variation in the variator's speed between no load to full load is 1% at the 1:1 ratio and within 3.5% at 4:1 reduction.
5. Drift under constant load is within 0.022% of output speed.
6. High efficiency of the drives is produced by the optimized rolling geometry of the rollers, disc, and ring; low loading of the rollers and bearings; and correct lubrication. Over the middle speed range, the efficiency is 93 to 84% at the extreme ratios. For units transmitting high power, this high efficiency results in substantial savings.
7. High output torque is obtained by transmitting the output power through the large diameter of the ring.
8. The roller variator may be operated with shafts in either horizontal or vertical position. It is adjustable for either direct drive or as an auxiliary control mechanism in the drive system. It operates in either direction of input shaft rotation.
9. Where high output speeds are required, the roller variator can be driven from the drive flange side, thereby multiplying the output speed up to five times the input speed.
10. Low output speeds as low as one-eighth of the input speed can be obtained and in many cases a reduction gear may be eliminated.
11. The speed of the variator cannot be changed while the unit is at rest, only while the unit is in motion.

Types of Traction Drives

Principles of operation. In the operation of the roller variator (Fig. 3.11), the flow of power is from the input shaft, through the pressure plate and drive disc, to the double conical rollers, which rotate on fixed inclined spindles. From the rollers, power is transmitted through the drive ring, pressure balls, and drive flange to the output shaft. The carrier in which the rollers are fitted can be moved axially by the rack and pinion.

The drive ring is partially immersed in an oil sump and lubricates all rotating parts. An oil pump is built-in in larger units between the bearings of the input side for positive oil circulation. The maximum ratio between low and high speed is 12:1. The output shaft rotates in opposite direction to the input shaft.

Figure 3.12 illustrates that the movement of the carrier increases or decreases the diameters at which the drive rollers are in contact with the drive disc and the drive ring, causing a change in output speed.

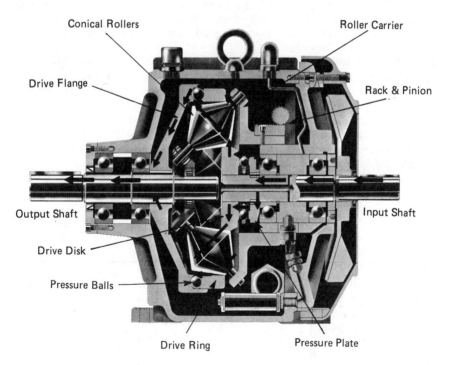

FIG. 3.11 Cutaway of the Kopp roller variator. (Courtesy of Koppers Company, Baltimore, Md.)

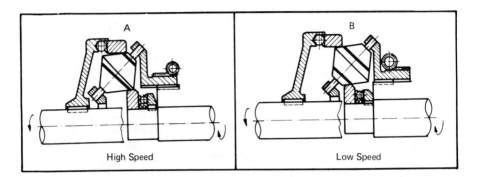

FIG. 3.12 Roller position and power flow at high and low speeds. (Courtesy of Koppers Company, Inc., Baltimore, Md.)

The pressure plate and pressure balls shown in Fig. 3.13 ensure that sufficient pressure, proportional to load demand, is applied to the rolling elements, thus providing positive traction whatever the load demand of the unit.

Torque-responsive mechanisms are incorporated on input and output sides, providing traction proportional to the torque transmitted through the drive. The pressure plate is keyed to the input shaft while the drive disc is free to revolve on this shaft. Both plate and disc have V-shaped ramps between which are retained steel balls.

As transmitted torque increases, the disc and plate move apart and the balls move up the V ramps to maintain an axial pressure on the drive disc and contact pressure between the conical rollers and the inclined face of the drive disc. The pressure is stronger than that required to transmit the power to protect against slippage. An identical responsive mechanism is between the drive ring and the drive flange.

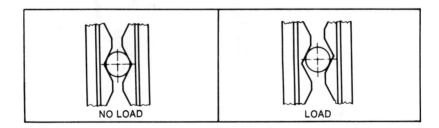

FIG. 3.13 Torque-sensing mechanism of the Kopp roller variator. (Courtesy of Koppers Company, Inc., Baltimore, Md.)

Types of Traction Drives

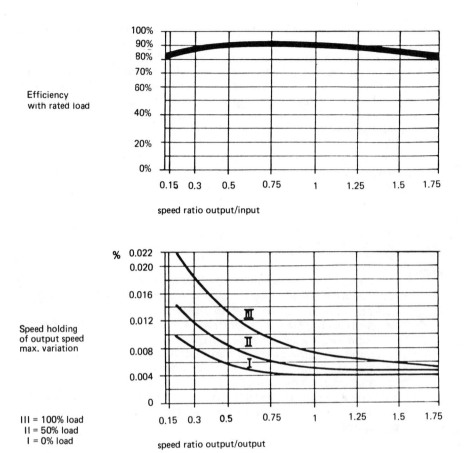

FIG. 3.14 Efficiency and speed-holding curves of the Kopp roller variator. (Courtesy of Kopp Variatoren AG, Mürten, Switzerland.)

The efficiency of a typical roller variator is plotted in Fig. 3.14. In this case, the curve is representative of a standard roller variator operated at 1800-rpm input speed with output speeds varying from 200 to 3000 rpm. The maximum efficiencies are realized when the drive is operated at the middle ratio ranges.

3.1.2 Ball Disc/Roller Disc

This type of traction drive consists of two versions: the disc-ball-disc and the disc-roller-disc.

The multiball disc drive, which was developed and patented by B. Rouverol of the Rolling Contact Gear Company of Berkeley, California, has been marketed under license by Graham Transmissions, Inc., since 1964 and P.I.V. Antrieb Werner Reimers.

Several other firms manufacture ball disc drives. P.S.I., which was purchased by Warner Electric, developed a light-duty single-ball drive. This unit was later manufactured and marketed by International Power Products. Hans Heynau of Germany developed a single-ball drive in 1963. The Heynau unit will be discussed in the free-ball types in Section 3.1.4.

Roller disc drives are manufactured by F.U. (Fabrication Unicum). These units were originally developed in 1936 and manufactured by Unicum S.A. of Paris, France. Tens of thousands have since been produced. In 1972, Parker Industries was appointed as sole U.S. agent.

Ball Disc

Operating features

1. The output speed range is adjustable down to 0 rpm.
2. The drive is adjustable while stationary or while running.
3. The unit is not sensitive to shock loads.
4. No tracking wear is experienced when operating at a fixed ratio setting for an extended time.
5. No starting or overload slip clutch is required.
6. The starting torque is 130% of running torque.
7. The drives are rated for constant output torque.
8. The drives are available for an output speed range of 1 to 10, or for an output speed from 0 rpm to 1.5 times the input speed.
9. The drives will operate in either direction of rotation. Depending on the type of unit, the input and output shafts rotate in the same or in the opposite direction to the input shaft.
10. Eight sizes are available, in capacities from 0.33 to 4 hp.

Principles of operation. The ball disc drive provides infinitely variable output speeds down to 0 rpm by employing balls as the power transmitting element. Both the input and output shafts (Fig. 3.15) carry a flat disc. A rotating cage of steel balls rolls between the two steel discs. The torque is transmitted through the balls from the input shaft to the output shaft. The ratio setting is determined by the position of the ball cage, which is adjusted by means of a control drive.

The balls, which are guided by the cage at all ratio settings, roll on the plane discs without slip. At each rotation the balls roll over changing paths. This prevents tracking wear even when the drive is operated at a fixed ratio for an unlimited period. Belleville springs distribute the required thrust loading.

Types of Traction Drives

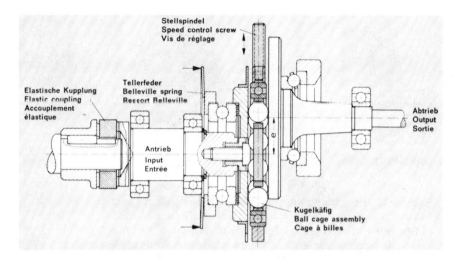

FIG. 3.15 Schematic of the P.I.V. ball disc drive. (Courtesy of P.I.V. Antrieb Werner Reimers KG, Bad Homburg, West Germany.)

Because of the infinite number of possible ratio settings, all imaginable forms of cycloids are produced by the freely rotating balls. Figure 3.16 shows the orbit lines on which the balls roll at ratio settings of 0.3, 1, and 1.2. The dark line indicates the orbit of one ball at one revolution. Figure 3.17 also shows the orbit lines in relation to the two discs.

When an imaginary center ball (representing the axis of rotation of the ball cage) is equidistant radially between the axis of the input

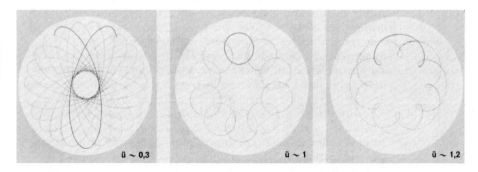

FIG. 3.16 Ball orbit lines at ratios of 0.3:1, 1:1, and 1.2:1. (Courtesy of P.I.V. Antrieb Werner Reimers KG, Bad Homburg, West Germany.)

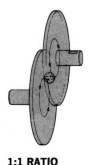 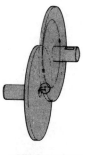 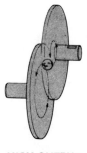

1:1 RATIO **LOW OUTPUT SPEED** **HIGH OUTPUT SPEED**

FIG. 3.17 Imaginary center ball position with respect to the input and output discs. (Courtesy of Graham Transmission, Inc., Division of Stowell Industries, Inc., Menomonee Falls, Wis.)

and output discs, a 1:1 ratio is obtained. Rotation of the input disc imparts spin to this imaginary ball, the surface velocity of which is the same as the velocity of the output disc at the point of contact.

Low output speeds occur when this imaginary ball is closest to the input disc axis and farthest from the output disc axis. The converse is true for high speeds.

Roller Disc

Operating features

1. The required output speed may be selected very precisely. The automatic pressure regulation compensates for varying load conditions. Speed variations are as low as 0.3%.
2. The efficiency will vary with output speed. The single-roller models are 77 to 92% efficient, and the four-roller models are 87 to 93% over the speed range.
3. The drives are rated for constant horsepower.
4. Speed ratios are available from 2:1 to 10:1.
5. To ensure efficient splash lubrication, the standard roller disc drive must be mounted upright with the shafts horizontal. Special versions are available for other mounting positions.
6. The roller disc drives are available in 1/3 to 100 hp ratings.
7. The speed should be adjusted only when the drive is running.
8. In one-roller models, the input and the output rotate in the same direction; in four-roller units they rotate opposite.
9. The units can operate either clockwise or counterclockwise.

Types of Traction Drives 45

Principles of operation. The roller disc variable-speed drives use double cone rollers rotating between input and output plates to transform a constant speed to a variable speed. One (Fig. 3.18) or four (Fig. 3.19) rollers are utilized, depending on the horsepower to be transmitted. Speed changes are achieved by moving the position of the rollers relative to the centers of rotation of the input and output plates. Figure 3.20 shows the path of power transmission in the one- and four-roller drives, the small arrows showing the power flow.

Positive traction and constant speed under varying load conditions are maintained by an automatic pressure system. This system comprises mating cam surfaces in the output shaft, which act on each other as the load increases to provide increased pressure between plates and rollers. This increased pressure is illustrated by the large arrows.

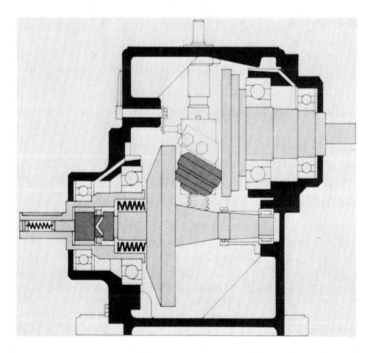

FIG. 3.18 Cutaway of the F.U. single-roller disc drive. [Courtesy of F.U. (London) Ltd., Leicester, England.]

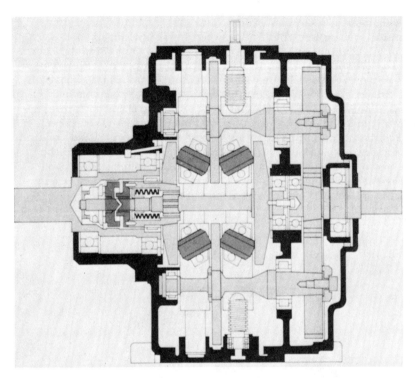

FIG. 3.19 Cutaway of the F.U. four-roller disc drive. [Courtesy of F.U. (London) Ltd., Leicester, England.]

3.1.3 Ring-Cone

The ring-cone traction drive has by far the largest configuration selection in the market. The ring-cone was the first variable-speed traction drive available to American industry.

Graham Transmissions, Inc., a division of Stowell Industries, Inc., introduced their drive in 1935 and it is still in production. The Graham ring-cone drive (Fig. 3.21) consists of a stationary metal ring with internally revolving tapered metal rollers. The ring is moved along the tapered rollers in order to obtain a ratio change. The input shaft turns the roller carrier. The rollers carry planet gears, which mate with a planetary ring gear that drives the output shaft. With this planetary feature, the output can be adjusted through zero and reverse.

Both Graham and Shimpo have a ring-cone drive (Fig. 3.24) that has an input sun wheel driving planet cones that are attached to

Types of Traction Drives   47

the cone carrier-output shaft. The inside diameter of the reaction ring moves back and forth on the outside of the cones, thereby changing the output speed. Shimpo features a dual-drive model with the input and output cones tied to a common reaction ring.

Shimpo also produces a simple ring cone (Fig. 3.33) without a planetary, with the speed ratio being controlled by changing the distance between the parallel input and output shafts. The input shaft moves relative to the fixed output shaft.

Another Shimpo unit (Fig. 3.30) consists of two parallel tapered roller cones connected by a movable rotating ring. The input and output shafts are offset and the movable ring is moved back and forth over the cones to change the speed ratio.

Hans Heynau of Germany developed a ring-cone traction drive (Fig. 3.36) in 1936. It is an all-metal drive similar to an adjustable sheave V-belt, but with a solid steel ring instead of the normal V-belt. When the load is applied, the ring transmits the torque as it is pulled into the steel cones. The speed ratio is obtained by varying the pitch diameter of the input and output cones.

Vadetec Corporation has modified the basic ring-cone principle with the addition of a nutating double cone (Fig. 3.37). Work on this

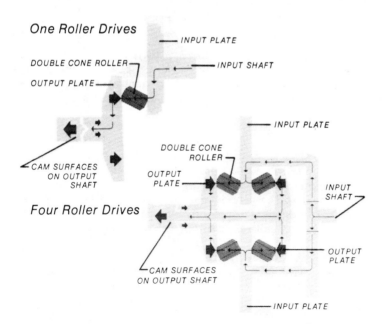

FIG. 3.20 Power flow for the F.U. single- and four-roller drives. (Courtesy of Parker Industries, Ind., Bohemia, N.Y.)

drive started in 1971. The unit's input shaft is attached to a rotating cylinder into which is set a free-turning double cone roller. As the input shaft rotates, the cone roller is forced to rotate against the stationary (nonrotating) adjustable traction rings. The cone roller's output end drives a planetary gear, which in turn drives the output shaft. The speed ratio is varied by moving the traction rings back and forth over the cones at different radii to obtain the desired output speed.

Ring-Cone with Planetary

Operating features. Special features of the Graham planetary gear ring-cone traction drive include:

1. Unlimited speed range from maximum speed to zero, and reverse. Speed is changeable running or stationary.
2. The drive will hold speed to within 0.1 to 1% of the maximum speed under steady loads.
3. The Graham drive protects itself and the driven machine against overload damage. Internal loading and tractive output is fixed.
4. Units are available in 12 sizes from 1/2 hp up to 5 hp.
5. The standard unit's input and output shaft rotate in the same direction.

Principles of operation. The Graham adjustable-speed drive (Figs. 3.21 and 3.22) uses a compound planetary gear system, except that the nonrotating member is a control ring which engages tapered rollers at varying diameters. The control ring is moved lengthwise backward and forward along the surface of the tapered rollers to change the speed ratio.

The input shaft supports the carrier for the three tapered rollers. The tapered rollers are inclined at an angle equal to their taper so that their outer edge is maintained parallel to the central axis. The required traction pressure between tapered rollers and the control ring is obtained through the centrifugal force of the rollers themselves, which press outwardly against the control ring under the centrifugal force. The pinion at the end of each tapered roller meshes with the ring gear, which is a part of the output shaft.

The pinions, which move with the carrier, turn at constant input speed around the central axis. Because of the traction between the rollers and the control ring, they must at the same time turn about their own axis in the opposite direction at a variable speed. This differential feature provides a maximum speed at the output shaft which is approximately one-third of the input speed.

Types of Traction Drives   49

FIG. 3.21 Cutaway of the Graham ring-cone drive with planetary. (Courtesy of Graham Transmissions, Inc., Division of Stowell Industries, Inc., Menomonee Falls, Wis.)

An infinite range of speeds can be obtained within the predetermined speed range of any particular drive. With the control ring at the large end of the tapered rollers, maximum speed is obtained, and with the ring at the small end, zero speed. A reverse speed feature can be incorporated into the transmission so that zero speed will occur at the midpoint of the rollers. Therefore, forward speed would be

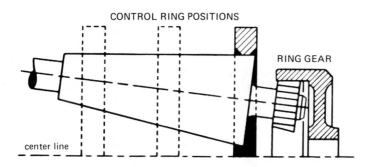

FIG. 3.22 Control ring positions for various ratios on the Graham ring-cone with planetary. (Courtesy of Graham Transmissions, Inc., Division of Stowell Industries, Inc., Menomonee Falls, Wis.)

obtained on the larger half of the rollers while reverse speed would be obtained at the smaller end of the rollers. The speed and direction of rotation of the output shaft is controlled by the ring gear-to-pinion ratio and control ring-to-roller diameter at the point of contact. By varying the pinion-to-ring gear ratio, various types of reversing and nonreversing drives can be obtained. The tapered rollers are supported by preloaded precision ball bearings.

Graham drives may be run in either direction of input. Standard transmissions are designed for fixed input speed and normally are used with built-in or coupled 3450-, 1750-, or 1150-rpm motors. Lower input speeds may be used, but output speed is directly proportional to the input speed. To find the maximum torque available from centrifugally loaded transmissions driven at less than rated input speed, multiply the catalog torque rating by the

$$\left(\frac{\text{driven speed}}{\text{rated input speed}} \right)^2$$

When operating at low or variable input speeds, the transmission should usually be spring loaded.

As the control ring is moved from the large end of the rollers to the small end, the leverage force is increased. The tractional or torque capacity of the Graham drive is thereby automatically increased at the smaller end of the rollers. In the standard drive, this produces increased torques at low speed—usually the most desirable condition.

The close-speed-holding feature of the Graham is shown in Fig. 3.23. Note how little the output speed varies with load at any given speed setting. Within the rating of the drive this close speed holding is comparable to the induction motor itself. The curves plotted here show that the Graham comes as close to a positive drive as is possible with an adjustable-speed transmission.

On drives without built-in gearboxes or with two-stage spur reduction, the output shaft rotates in the same direction as the driving motor. On drives with built-in single spur reduction, the output shaft rotates in the opposite direction of the motor. Direction of rotation on drives with built-in worm reduction depends on the arrangement selected. The opposite of the statements above is true when the drive selected has the characteristics of increased torque at high speeds.

All units should be mounted horizontally. Special modifications are available for vertical, side, or other special mounting of most models. Starting torque ratings are approximately 75% of the running torque for a centrifugally loaded unit, and approximately 85% of the running torque for a spring-loaded unit.

If input speed is lower than standard or variable, spring-loaded

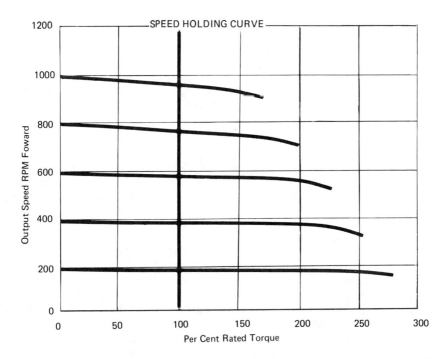

FIG. 3.23 Speed-holding curve for the Graham drive. (Courtesy of Graham Transmissions, Inc., Division of Stowell Industries, Inc., Menomonee Falls, Wis.)

transmissions are used. The tapered rollers are held against the control ring by spring pressure rather than by centrifugal force, as in drives with higher input speeds. When spring loaded, the drives should only be shifted while running.

The differential Graham drive is designed so that a complete stall of the output shaft will not cause a complete skidding of the tractive parts. Even if the driven unit stops completely and the variable-speed drive is set for maximum speed, over 75% of the roller movement is rolling and the relative motion, or sliding, is only 25%. A new roller surface contact is continually being offered, new lubricant is constantly drawn in, and the pressure, instead of building up excessively, is limited by the centrifugal force.

Single-Sun Wheel Ring-Cone

The Graham and Shimpo sun wheel ring-cone traction drive are very similar units with the exception that the Graham units are available in sizes up to 20 hp versus only 5 hp for the Shimpo units.

Operating features

1. The drives are compatible with any standard C face motor for minimal overall length.
2. The drives can sustain a 225% starting torque. They can provide steady output speed under fluctuating load. They have the ability to absorb shock loads.
3. The drive can be used as a constant-horsepower speed variator in certain speed variation ranges and as a constant-torque type in all the ranges.
4. The drive has a 4:1 speed range and an inherent 4:1 speed reduction (3:1 on 15- and 20-hp models). Therefore, output speed range is 1/4 to 1/16 of the input speed (1/3 to 1/12 on 15- and 20-hp models).
5. Speed reducers offering an additional reduction are available.
6. The units can be flange or foot mounted.
7. All units are for horizontal mounting unless otherwise specified.
8. Input rotation may be in either direction of rotation.
9. The output rotation of the drive with planetary reducer, or with double-reduction Circulute reducer, as in the same direction as the input. Output rotation of the drive with single-reduction Circulute is the opposite of input rotation.
10. The unit can be adjusted only while in motion.

Principles of operation. The single-sun wheel ring-cone drive (Fig. 3.24A) employs a planetary system. A sun wheel (6) mounted on the input shaft (7) contacts five planetary cones (2). These cones are mounted on five spindles and maintain contact with the sun wheel as well as the inner surface of the control ring (1). These spindles, mounted on the cone carrier (3), are angled to keep the outer surface of the cones parallel to the centerline. The cone carrier is connected to the output shaft (5) through a pressure control cam (4).

The five planetary cones are driven by traction and rotate on their own spindles. At the same time, the cones revolve around the inner surface of the stationary control ring. The revolving motion of the planetary cones is transmitted through the cone carrier and pressure control cam to the output shaft.

Although the control ring does not rotate, it is moved axially by the handwheel through a rack and pinion. When the ring is in contact with the large end of the planetary cones, the roller revolving speed (i.e., output speed) is maximum and is equivalent to 1/4 of the input speed. By moving the ring toward the smaller end of the cones, the output speed slows down. Minimum speed is one-sixteenth of the input speed. This is shown in Fig. 3.25. Therefore, with 1750-rpm input speed, turning the handwheel will provide an infinite speed change of the output shaft from 450 rpm to 112 rpm.

The pressure control device consists of two ramped cams and

Types of Traction Drives

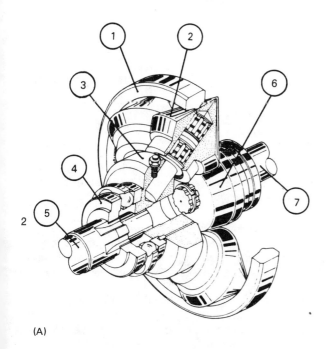

(A)

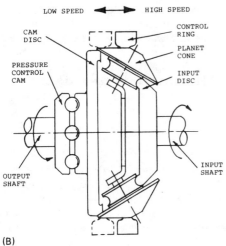

(B)

FIG. 3.24 (A) Cutaway of the single-sun wheel ring-cone. (B) Schematic of the 0 to 1000 rpm single-sun wheel ring-cone. (Courtesy of Graham Transmissions, Inc., Division of Stowell Industries, Inc., Menomonee Falls, Wis.)

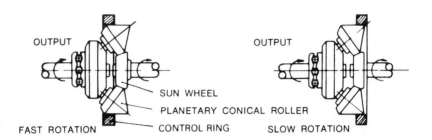

FIG. 3.25 Schematic of the ring and roller positions for fast and slow rotation. (Courtesy of Graham Transmissions, Inc., Division of Stowell Industries, Inc., Menomonee Falls, Wis.)

steel balls (Fig. 3.26). The cam loading applies contact pressure between the cones and the sun wheel/orbit ring combination. Since cam A does not move axially, as the load increases on the output shaft during operation, the steel balls climb up the V-shaped ramps and push cam B together with the cone carrier toward the input shaft. This results in an increase in contact pressure and avoids slippage on the contact surface. When the load decreases, the pressure is reduced and contact surfaces are prevented from being placed under excessively high pressure when it is not required at light loads.

Graham has introduced a new version of the single-sun wheel ring-cone traction drive that is very similar to the unit just described, except that the basic unit can operate with an output range of 0 to 1000 rpm with a 1750 rpm maximum input speed. The output torque increases 700% as the output speed is reduced from 1000 rpm to zero.

The principles of operation (Fig. 3.24B) are the same as for the original single-sun wheel ring-cone traction drive except that the planetary cones are set in a floating planet carrier rather than the carrier being connected to the output shaft through the pressure control cam.

The planet cones are driven through grooved discs by traction from the input disc on the input shaft and also rotate on their own axes on the floating planet carrier. The cone portion of the planet cone rotates on the inner surface of the nonrotating but axially movable control ring. The rotary motion of the cones is transmitted to the cam disc, which rotates on the underside of the planet cones. The cam disc is connected to the output shaft through the pressure control cam, which consists of two ramped cam plates and steel balls.

When the ring is in the high-speed position, the cones will rotate rapidly on their axes and revolve slowly around the inner surface of the control ring. This rapid cone rotation is transmitted to the cam disc, giving high output speed. Moving the control ring toward the low-speed position decreases the speed of rotation of the cones and increases their rate of revolution around the control ring, which decreases the output speed.

Types of Traction Drives 55

Zero output speed is achieved when the control ring contacts the planetary cones at the same diameter as the contact point of the cam disc.

The direction of rotation of the output shaft is opposite that of the input shaft except at zero when there is no output motion.

Dual-Sun Wheel Ring-Cone

Operating features. Shimpo's dual-cone-construction unit with a common ring between the cones has the following features in addition to the single-ring cone unit:

1. Forward and reverse variable rotation is available through all attainable speeds, from fast forward through zero rotation (i.e., output shaft at rest with the motor running) to full reverse.
2. Since driving from zero begins at the area of largest friction surface, torque is high. Constant stable rotation and sufficient torque can be expected over 30 rpm both in forward and reverse rotation. In this configuration, the speed ratio is 10:1 forward and 12:1 in reverse.
3. The unit is available in sizes from 1/8 to 5 hp.
4. The drive is a constant-horsepower device.
5. The unit can be mounted horizontally and vertically.
6. The speed can only be adjusted when the unit is running.

Principles of operation. The principle of power transmission is shown in Fig. 3.27. The main components are (1) two sun wheels which are splined to the motor shaft, (2) two sets of planet cones (one fixed and one connected to output shaft), and (3) a solid bridging ring. There are five planet cones in each of the two sets; only seven are shown in the diagram. The fixed planet cones are in contact with

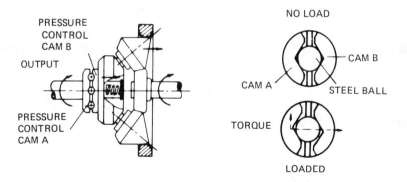

FIG. 3.26 Schematic of the pressure control device of the single-sun wheel ring-cone drive. (Courtesy of Graham Transmissions, Inc., Division of Stowell Industries, Inc., Menomonee Falls, Wis.)

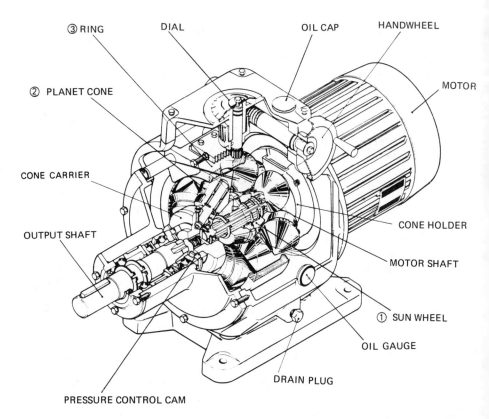

FIG. 3.27 Cutaway of the dual-sun wheel ring-cone drive. (Courtesy of Shimpo Industrial Co., Ltd., Kyoto, Japan.)

and driven by the sun wheel, rotate on their own axes, but do not revolve. The output shaft planet cones, also driven by the sun wheel, not only rotate on their individual axes, but also revolve in orbit around the sun wheel in the cone carrier. The cone carrier is connected to the output shaft through the pressure control cam. One contact face of the bridging ring is in contact with and driven by the fixed cones, and the other is in contact with the output shaft planet cones.

With the bridging ring positioned at the base (i.e., large diameter) of the output shaft planet cone, and the apex of the fixed cone (Fig. 3.28A), the output shaft rotates in the same direction and at one-fourth of the speed of the input shaft. The ring is stationary at this position. The planet cones, like the fixed cones, are driven by the sun wheel, which is in constant rotation. The planet cones also maintain constant traction on the contact face of the bridging ring. Thus the cone carrier revolves at a speed in proportion to the difference of surface travel between the apex and base of the cones.

With the ring positioned at midpoint (Fig. 3.28B), each contact face at the center point of its respective set of cones, the planet cones spin at the same speed as the fixed cones, but in the opposite direction. Thus the cone carrier and the output shaft are stationary. The output shaft maintains torque and will not respond to force in either direction exerted by the driven machinery.

Moving the ring in the direction of the fixed cones (Fig. 3.28C), the fast surface travel of the base of the driving cones rotates the ring at relatively high speed. The output shaft planet cones, due to the ring contact surface at the apex, are rotated (on their own apexes) and revolve around the sun wheel at a lower speed than the ring. As these planet cones are connected to the output shaft by the carrier, direction of rotation is now reversed and speed is one-third that of the input shaft.

The ring-cone traction drive protects itself and the driven machine against overload damage. This mechanism (Fig. 3.29) is identical to that for the single-ring cone unit. The pressure control mechanism has cams and steel balls. The cone carrier is pressed by the action of a "wedge." This applies pressure to the friction surface of the transmission. The amount of pressure automatically adjusts according to the load.

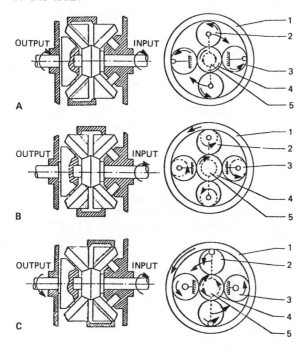

FIG. 3.28 Input and output shaft rotation with respect to ring roller position on the dual-sun wheel drive. 1, ring; 2, planet cone; 3, fixed cone; 4, sun wheel; 5, output shaft. (Courtesy of Shimpo Industrial Co., Ltd., Kyoto, Japan.)

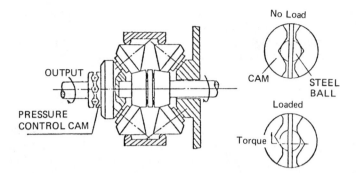

FIG. 3.29 Schematic of the pressure control cam of the dual-sun wheel drive. (Courtesy of Shimpo Industrial Co., Ltd., Kyoto, Japan.)

Parallel Cone

Another Shimpo ring-cone traction drive consists of two parallel cones, one for input and one for output, and a ring.

Operating features

1. The variator consists of a combination of two cone shafts and a ring. The speed ratio of the output to input shaft ranges from 1/2 to 2, so that an overall speed ratio of 4:1 is obtainable. For instance, if the input speed is 1200 rpm, the maximum output speed is 2400 rpm and the minimum speed is 600 rpm.
2. The drive is available in sizes from 1/16 to 3 hp.
3. The recommended input speed is from 900 to 1200 rpm.
4. The output shaft rotates in one direction according to pressure control mechanism and is not reversible.
5. This model is a constant-horsepower device.
6. The drive is for horizontal mounting only.
7. The speed can be adjusted only when the drive is operating.

Principles of operation. The main parts of this drive (Fig. 3.30) are composed of two cone shafts which are positioned face to face, and the steel rotating ring connecting them. The transmission ratio is varied by moving the ring along the cone shafts. If the ring is positioned at Fig. 3.31B, the ratio of input to output is 1:1; if it is at Fig. 3.31A or 3.31C, the ratio is 1:2 and 2:1, respectively.

The pressure control mechanism in Fig. 3.32 works as follows. When the roller (A) of the input shaft begins to rotate in the direction

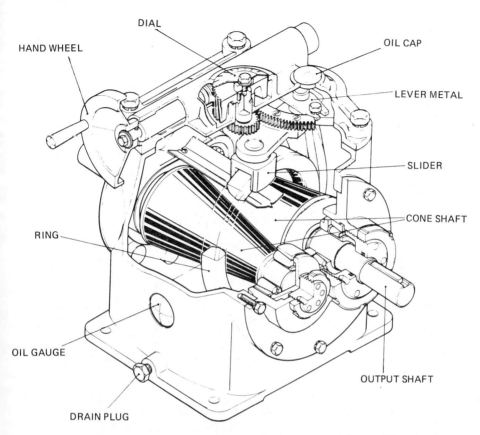

FIG. 3.30 Cutaway of the parallel-cone ring-cone drive. (Courtesy of Shimpo Industrial Co., Ltd., Kyoto, Japan.)

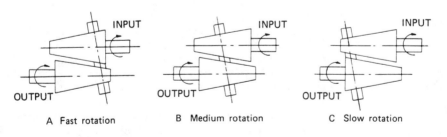

FIG. 3.31 Rotational speed versus ring and cone position of the parallel-cone drive. (Courtesy of Shimpo Industrial Co., Ltd., Kyoto, Japan.)

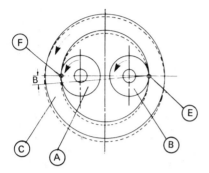

FIG. 3.32 Pressure control mechanism of the parallel-cone drive. (Courtesy of Shimpo Industrial Co., Ltd., Kyoto, Japan.)

shown by the arrow, the roller (B) of the output shaft is at rest. As the pressure of the ring (C) is increased, it transmits the moment of rotation. The ring (C) automatically shifts point (F) along the inside curvature of the ring with point (E) as a fulcrum. Thus internal contact pressure is automatically increased, but only during operation.

Variable-Position Input Shaft Ring-Cone

Shimpo manufactures a ring-cone traction drive which achieves speed ratio variation by changing the input shaft position in relation to the fixed-position output shaft.

Operating features

1. This model has a wide speed range of 10:1, and is suitable for high-speed operation.
2. The output speed can be preset or changed as desired whether the motor is running or is stopped.
3. This drive is manufactured in sizes from 1/16 to 2 hp.
4. This drive operates in either direction, clockwise or counterclockwise, and the input and the output shafts rotate in the same direction.
5. The drive can be used as a constant-horsepower variator in certain speed ranges and as a constant-torque type in all the ranges.
6. The drive can be mounted horizontally only.

Principles of operation. The drive (Fig. 3.33) is composed of the movable umbrella-shaped cone shaft and the disc-shaped ring. When the centerline of the cone shaft is changed relative to the ring,

Types of Traction Drives

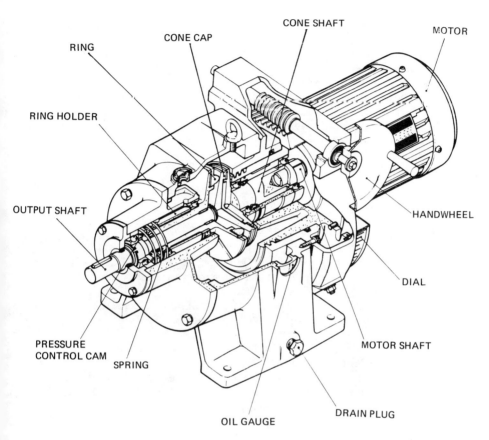

FIG. 3.33 Cutaway of the variable-position input shaft ring-cone drive. (Courtesy of Shimpo Industrial Co., Ltd., Kyoto, Japan.)

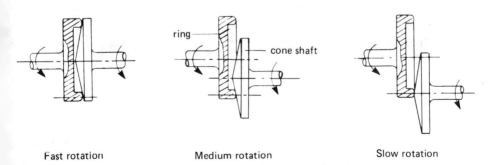

FIG. 3.34 Ring and cone position versus output speed. (Courtesy of Shimpo Industrial Co., Ltd., Kyoto, Japan.)

contact position varies, providing a 10:1 stepless speed ratio (Fig. 3.34).

The automatic pressure control mechanism in Fig. 3.35 consists of steel balls between two cams. In the no-load condition, the steel balls are in the bottoms of the grooves. As the output shaft load is increased, the wedging action causes pressure in the axial direction. This moves the ring holder axially and acts on the frictional surface of the transmission giving positive, nonslip traction. Pressure decreases automatically as the load is reduced.

Variable-Pitch Diameter Cone and Ring

The Hans Heynau ring-cone traction drive consists of two variable-pitch parallel offset cones connected by a steel ring.

Operating features

1. The drive is a constant-horsepower design, but can be used as constant torque at maximum output speed value over entire adjustment range.
2. The standard 6:1 or 9:1 speed range is 3:1 to 1:2 or 3:1 to 1:3.
3. Output speeds up to 5400 rpm can be attained.
4. The unit is 85% efficient.
5. The accuracy is ±0.1% of output speed at constant load.
6. The unit is available from 1/3 to 7.5 hp.
7. Speed adjustment is permissible at rest or while running.
8. Infinitely variable adjustment range 0 to 750 rpm is available through differential action.

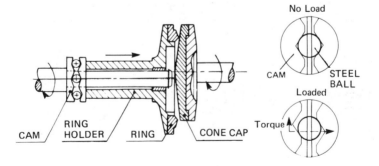

FIG. 3.35 Schematic of the pressure loading mechanism of the variable-position input shaft ring-cone drive. (Courtesy of Shimpo Industrial Co., Ltd., Kyoto, Japan.)

Types of Traction Drives 63

9. Input to the unit is indicated by cast in arrows, but the drive may be operated in either direction of rotation.
10. The contact pressure between transmission parts is proportional to the load torque.
11. It can be supplied with or without motor, C flange, or coupled versions.
12. The output rotation is in the same direction as the input.
13. The input and output shafts are offset but parallel.

Principles of operation. The significant components of the Heynau drive (Fig. 3.36) are two pairs of cones [(1 and 2), (3 and 4)] mounted on the input and output shafts and connected under tension by a solid steel ring (5). This ring, which has precision ground contours, rides on the surface of the four cones to transmit power from the input shaft to the output shaft.

Two of the cones, diagonally opposed to each other (1 and 4) are rigidly fixed in the housing while the other two (2 and 3) are arranged to slide on splines, thus permitting the ring (5) to revolve on different adjustable cone diameters. The adjustment provides a speed ratio between the input and output shafts of 1:3 (maximum speed) and 3:1 (minimum speed)—in other words, a total variation of 1:9.

The construction of the drive with an adjustment range of 1:6 is identical except that the upper output speed range is limited by reducing the movement of the two adjustable cones so that there remains a variation of 3:1 and 1:2 (i.e., a total variation of 1:6).

The two movable cones slide in guide bearings (6 and 7) which are rigidly connected by two tie rods (8). The contact pressure necessary for producing the required friction between ring and cones is adjusted with the threaded ring (9).

Where a load is applied, the transmission ring is pulled into the triangle formed by the cone pairs within the range of the parts under load. The ring movement, and the contact pressure, is directly proportional to the output torque. The contact pressure automatically adjusts to meet the load applied.

Nutating Cone and Ring

Operating features. The operating features of the Vadetec nutating traction drive include:

1. The speed range extends from 0 to 1:1.
2. There are provisions for output gear set arrangements.
3. Efficiencies reaching 90 to 94% are attainable.
4. Axial forces from the two contact areas cancel each other, eliminating heavy thrust loads.
5. The drive is available in sizes from 15 to 150 hp.

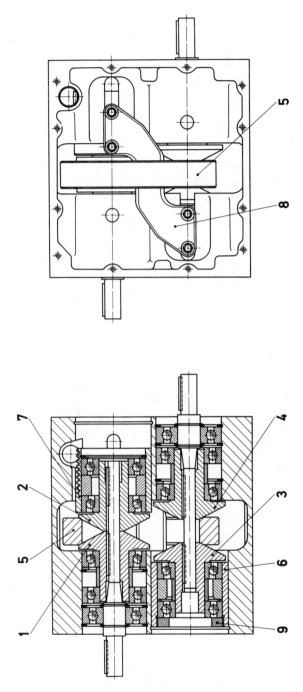

FIG. 3.36 Schematic of the Heynau H-drive ring cone. (Courtesy of Plessey Dynamics Corp., Hillside, N.J.)

64

Types of Traction Drives 65

    6. The ball ramp method of loading the contact areas allows rapid response to variations of torque.
    7. It can be rotated in either direction of rotation.
    8. Either shaft can act as the input or output.
    9. Output direction is opposite the input direction on the basic unit.
    10. The speed can be regulated either during operation or when stationary.
    11. A drive that is reversible through zero is being tested.

Principles of operation. The Vadetec nutating traction drive is composed of five basic components shown in Fig. 3.37.

    1. The transmission case or reaction member
    2. The input or driving member
    3. The nutating or rolling assembly
    4. Two control rings or reaction members
    5. The output or driven member

As viewed from the right end of this unit, the input torque to the driving member is in a clockwise direction.

    The driving member casting has a cylindrical central section which has been hollowed out to accommodate the nutating biconical roller assembly (see Fig. 3.37). The shaft of the nutating assembly, which is mounted on bearings in the input member, is skewed at an angle so that its axis intersects the axis of the input member at the center. As the input member rotates, the nutating shaft nutates; that is, its axis describes two cones with their apexes at the point of intersection of its axis with the axis of the input member.

    Mounted on the nutating shaft are two central collars, which are splined to the shaft, and two cones, which are free to both slide and rotate on the shaft. On one facet of each central collar and on the base of each cone are two kidney-shaped grooves of semicircular cross section, which are sloped so that one end is deeper than the other to form a ramp angle, as shown in Fig. 3.38A. When mated, the grooves form pockets, each containing a ball so that the cones are urged outward by these ball ramps when twisted on the nutating shaft. The ball ramps of one cone are indexed at 90° with respect to the other so as to provide a universal joint action. Between the two collars, a Belleville spring maintains a preload on the system. This loading configuration has been changed on the later models.

    Now on the large end of the outer right-hand cone and on the face of a flange on the inner cone shaft are three kidney-shaped grooves of semicircular cross section, which are sloped so that one end is deeper than the other to form a ramp, as shown in Fig. 3.38B. When

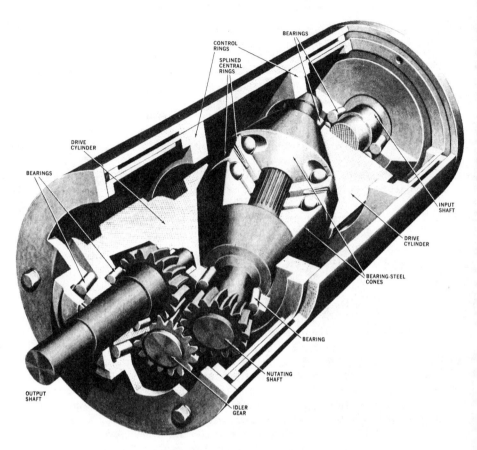

FIG. 3.37 Cutaway of the Vadetec nutating cone drive. (Courtesy of *Popular Science*, Times Mirror Magazines, Inc., New York.)

mated, each pair of grooves becomes a pocket and in each pocket is a ball. The ball ramps thus formed urge the two roller cones apart when twisted with respect to each other. This is possible because the needle bearings separating the cones and the roller bearings supporting them permit axial freedom as well as rotational movement. Behind the cone shaft flange is a set of Belleville springs which provides a preload on the system.

The two cones are in contact with two control rings through the openings in the sidewall of the input member. One of these points of contact is at the lower left of the left-hand cone; the other is at the upper right of the right-hand cone, as shown in Fig. 3.37. The

Types of Traction Drives 67

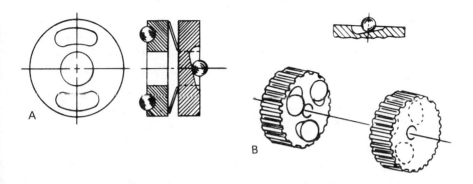

FIG. 3.38 (A) Torque-loading mechanism of the early nutating drive. (B) Refined torque-loading mechanism. (Courtesy of Vadetec Corporation, Troy, Mich.)

nutation of the cone and shaft assembly causes the cones to roll on the inside periphery of the control rings in an epicyclic or planetary motion, but in a direction opposite to input rotation.

The tangential force developed at each of the two contact points tends to twist the cones on the nutating shaft, thereby causing the entrapped balls to roll up the ramps and force the cones outward, which wedges them into the control rings. Thus the contact pressure is a function of the torque transmitted and, by appropriate design of the ball ramp angle, is always maintained at the level necessary to provide the required traction. The output planetary motion of the nutating shaft assembly can be viewed as orbiting about the main axis, at constant input speed and direction, while rotating about its own axis, or spinning, at speeds greater than input and in the opposite direction.

The rotational speed of the shaft about its own axis is a function of the geometric ratio, represented by the perpendicular distance from the axis of the nutating shaft to the contact point, to the perpendicular distance from the axis of the input member to the contact point. The control rings are keyed to the transmission case, which is the reaction member, and can slide axially toward either the center or the ends of the case. Thus, by varying the position of the rings (Fig. 3.39), the points of contact are shifted so that the rotational speed, and the speed ratio, are changed in accordance with the geometric ratio.

If the control rings are at the center of the case and the diameter of the base of the cones is, for example, close to the inside diameter of the control rings, the speed ratio is approximately zero (because the cones just wobble within the control rings without being able to roll) and the torque is maximum.

If the control rings are at the ends of the case and the diameter of the small ends of the cones is, for example, equal to half the inside diameter of the control rings, the speed ratio is 1. For this ratio of diameters, the nutating shaft is rotating at twice input speed, but in the opposite direction.

In another arrangement, the drive pinion (Fig. 3.40B) on the end of the nutating biconical assembly meshes directly with an internal face gear mounted on the output shaft in place of the sun gear arrangement (Fig. 3.40A). Instead of being additive, the relative effects of the rotational and orbital speeds are subtractive, and now the low and high ends of the ratio range are reversed.

Minimum output speed occurs when the control rings are at the extreme outer position, because in this position the rotational speed of the nutating biconical assembly is at its maximum' and is subtracting the most from its orbital speed. Maximum output speed occurs when the control rings are close together because in this position the rotational speed of the nutating assembly is at its minimum, slowing down and subtracting less and less from its orbital speed.

This type of output gear set arrangement is capable of providing underdrive output speeds with precise speed control. Through the appropriate selection of the gear ratio, it is possible to reach a zero point within the range and even pass through zero so that both forward and reverse speeds are provided.

Since the output end of the nutating shaft orbits about the main axis of the transmission, it is necessary to bring its motion back into line with the main axis. This is accomplished through various output gear set arrangements, depending on the specific application.

3.1.4 Free-Ball

The Contraves free-ball traction drive was developed by Martin Arter in Switzerland in 1963. Over 30,000 units are in operation in Europe and the United States. There are two versions of the Contraves free-ball drive, the difference being the final output range. One version has a range from 0 to 80% of input speed with zero in the midpoint with direction change on either side. The other gives up to 20:1 overall ratio with no direction change unless the power source is reversed. The Contraves units use four free balls and control rollers with the driver and the driven discs.

Floyd Drives Company of the United States manufactures a free-ball drive similar in principle to the Contraves, but using a planetary system in conjunction with two balls and control rollers and driver and driven discs. Hans Heynau of Germany began manufacturing a free-

Types of Traction Drives

Maximum torque, zero speed

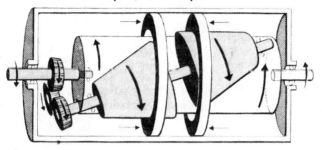

Minimum torque, maximum speed

FIG. 3.39 Torque versus speed relationship of the nutating cone drive. (Courtesy of *Popular Science*, Times Mirror Magazines, Inc., New York.)

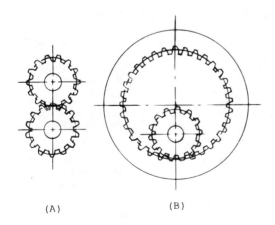

(A)   (B)

FIG. 3.40 Sun-planet and internal gear output configurations for the nutating cone drive. (Courtesy of Vadetec Corporation, Troy, Mich.)

ball unit in 1963. This unit could also be considered under Section 3.1.2, "Ball Disc," but is discussed with the free-ball units because of the manner of control of the single ball between the input and output discs.

Bidirectional Four-Ball and Roller Free-Ball

The Contraves bidirectional free-ball unit consists basically of an input disc, four free balls, four control rollers, and an output disc.

Operating features

1. The speed can be regulated either during operation or when stationary.
2. The control spindle can be mounted on the driving or driven shaft side.
3. Accurate repeatability of the desired speed is obtainable.
4. The input and output drive shafts are coaxial.
5. The ball pressure varies with output torque.
6. The unit is bidirectional from 0 to 80% of input speed with zero speed being the midpoint setting.
7. The overall efficiency is 80 to 90%.
8. The speed change at load is less than 1%.
9. The drive is available in sizes from 0.25 to 2 hp.
10. The drive can be accelerated at a high rate of speed.

Principles of operation. Torque is transferred from the driving shaft (Fig. 3.41) to the driven shaft by the stationary ball race, four balls, the output-side ball race, and the thrust assembly. The driving shaft is supported in the variator housing and roller carrier. The large thrust collar is fixed radially to the driving shaft, but can be axially displaced on it. The small thrust collar is rigidly connected to the flange on the output-side ball race. Rubber rings are inserted between the thrust collar, flange, and ball race.

Three individual grooves are ground on the front of the thrust collars and decrease in depth in both directions of rotation. When the speed variator is in the no-load status, the balls remain at the deepest point of the grooves. The springs produce the initial thrust pressure. The thrust collars turn in opposite directions with increasing torque while the pressure balls climb along the groove slope.

The changing contact pressure produced in this manner is proportional to the torque. It is transmitted to the ball races, balls, and control rollers. The centers of the balls and the control rollers are in one plane and form a square. The ball centers are in the corner points of the square and the control roller centers are in the middle of the side lines.

Types of Traction Drives 71

FIG. 3.41 Cutaway of the Contraves type 0-1000 free ball drive. (Courtesy of Contraves AG, Zürich, Switzerland.)

The control rollers are incorporated in the roller shift wheels, which, in turn, are supported in the roller carrier, which is rigidly connected to the drive shaft and turns with it. The control worm, which is connected to the control spindle, engages in the toothed rims of the roller shift wheels. If the control spindle is axially shifted, the four roller shift wheels rotate with their control rollers. The theoretical rotary axes of the control rollers and the steel balls always intersect at one point on the theoretical axis of the speed variator axis. With the axes parallel the rolling circles are equal and the driven shaft stops. If the control roller axles incline toward the driven shaft, it rotates in the opposite direction to the driving shaft. If the control roller axles are inclined toward the driving shaft, the driving and driven shafts both rotate in the same direction. The more the control roller axles are inclined toward the variator axis, the greater the output speed.

Four-Ball and Roller Free-Ball

The second Contraves free-ball unit contains the same major components but is not bidirectional unless the power source is reversed.

Operating features

1. The speed can be regulated either during operation or when at a standstill.
2. The control spindle can be mounted at the driving- or the driven-shaft side.

3. Accurate repeatability of the desired speed is obtainable.
4. The driving and driven shafts are coaxial.
5. The ball pressure varies with output torque.
6. The drive can be operated in either rotational direction.
7. The overall efficiency is 85 to 90%.
8. The speed change at load is less than 1%. The drive is available in sizes from 0.2 to 5 hp.
9. Available ratios are up to 20:1 (i.e., 10:1 to 1:2).
10. The drive can be accelerated at high rate of speed.

Principles of operation. The torque is transmitted from the driving shaft (Fig. 3.42) by the driving-side thrust assembly, the driving-side ball race, the four steel balls, the driven-side ball race, and the driven-side thrust assembly to the driven shaft. The control spindle is rigidly connected to the control worm and cannot be moved axially. The control worm is in mesh with the four roller shift wheels with external teeth, which in turn are supported in the stationary roller carrier held by the roller carrier disc.

The centers of the roller shift wheels, control rollers, and steel balls are situated in one plane and are stationary. The steel balls are centered by the four control rollers supported in the roller shift wheels, and the two ball races. The two thrust assemblies press the ball races against the steel balls.

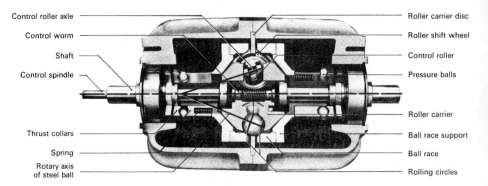

FIG. 3.42 Cutaway of the Contraves standard-type free-ball drive. (Courtesy of Contraves AG, Zürich, Switzerland.)

Types of Traction Drives 73

The thrust assemblies consist of two thrust collars and three pressure balls with springs providing the basic ball pressure. The broad thrust collars are rigidly centered on the two ball race supports, but are axially movable. These supports may rotate to a limited extent on the driving and driven shafts, respectively.

The driving and driven shafts are both supported in the roller carrier and in the variator housing. The narrow thrust collars are rigidly mounted on the two shafts, but are axially movable as well. Each thrust collar carries three individual grooves for the pressure balls, with decreasing depth in both directions of rotation. As long as the speed variator is on no-load, the balls remain at the deepest point of the grooves. With increasing driving torque, the pressure balls climb along the groove slope, exerting an axial force proportional to the torque.

Turning the control spindle will vary the inclination of the control roller axles, resulting in a variation of the speed of the driven shaft. The inclination of the rotary axes of the steel balls is determined by the inclination of the rotary axes of the control rollers. This fact permits a variation of the perimeter ratio of the driving- and driven-side rolling circles on the steel balls.

The theoretical rotary axes of the control rollers and the steel balls always intersect at one point on the theoretical axis of the speed variator. However, if the control roller axles are parallel to the variator axis, the output speed will be identical to the input speed, the two rolling circles being equal.

If the control roller axles are inclined toward the axis of the driving shaft, the output speed will decrease with increasing inclination of the control roller axles, the driving-side rolling circle growing larger and that on the driven side becoming smaller. If the control roller axles are inclined toward the axis of the driven shaft, the output speed will increase in relation to the input speed.

Single-Ball Free-Ball

The Heynau free-ball traction drive unit consists of an input disc or cone, a single adjustable free ball, and an output disc or cone.

Operating features

1. The drive is a constant-horsepower device, but can be used as constant torque at maximum output speeds over the entire adjustment range.
2. Standard speed ratios are 6:1 or 9:1 (i.e., 3:1 to 1:2 and 3:1 to 1:3).
3. The output speeds can be up to 5400 rpm.

4. The drive is 85% efficient.
5. Speed accuracy is ±0.1% of output speed at constant load.
6. The drive is available in sizes from 1/8 to 1/2 hp.
7. Speed adjustment is permissible at rest or while running.
8. The unit can be run in either direction of rotation.
9. The units can be supplied with or without motor, C flange, or coupled versions.
10. Gearbox ratio selections are available.
11. The output rotation is the same direction as input.
12. The input and output shafts are offset but parallel.

Principles of operation. The major components of the Heynau drive (Fig. 3.43) are steel ball (1) positioned between two axially displaced hollow cone discs (2 and 3) and acts as a power transmission element. When the load is applied, the transmission ball is pulled into a triangle formed by the two hollow cone discs by an amount equal to

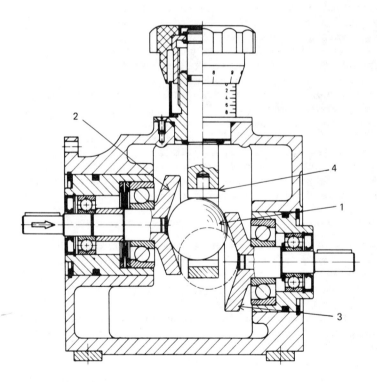

FIG. 3.43 Schematic of the Heynau mini-drive. (Courtesy of Plessey Dynamics Corp., Hillside, N.J.)

Types of Traction Drives

the elastic deformation of the parts under load. Thus the contact pressure is directly proportional to the output torque. Torque-dependent pressure devices are unnecessary.

Clockwise or counterclockwise rotation is permissible. The output speed of the Heynau drive is infinitely variable and is achieved by adjusting the position of the steel ball by rotating the speed-setting spindle knob (4). Speed setting is permissible both at rest and in motion.

In the upper adjustment position, a ratio of 3:1 reduction is created between input and output shaft. In the lower adjustment position the ratio is a 1:3 speed increaser. The total speed range covered is 9:1. For a speed range of 6:1, higher input horsepower is possible since the output horsepower is determined by the lower output speed.

The movement of the transmission ball is positively controlled when adjusting for high and low speeds. When adjusting to the lower speeds, the transmission ball takes up a position against the speed-setting spindle because of its tendency to move toward the middle of the higher cones. Power must be transmitted through the unit only in the direction shown by the arrow on the outer housing. In the case of very low input speeds, a minimum amount of load must be applied at the output shaft to achieve the desired output speed. The drive may be used in any mounting position and is hermetically sealed.

Two-Ball and Control Roller Free Ball

The Floyd free-ball drive consists of an input ring, two drive balls, two control rollers, radial rollers, a driven disc, an overload device, and a planetary gear output.

Operating features

1. Maximum torque is available at zero speed.
2. Full continuous rated torque is of 17.5 lb in. and is available at all speeds, including zero.
3. It has a temporary overload capability of approximately 70 lb in.
4. Full speed is available in either direction, 1800 rpm forward to 0 to 1800 rpm reverse.
5. The output speed is linear within 1% of the control shaft angle.
6. Speed variation with load is 3 to 5%.
7. The efficiency is 75%.
8. The drive is only available in 1 hp.
9. The speed ratio can be adjusted while the unit is at rest or while in motion.

Principles of operation. The Floyd free-ball traction drive is similar in principle to the Contraves drive in that it has axle-free balls guided by control rollers. One difference is that the Floyd unit has only two balls (Fig. 3.44) and incorporates a planetary gear system to gather up the torques.

Torque from the input ring is divided between the axial roller and the radial rollers, which drive into a disc plate. The disc plate drives through a cammed clutch into the ring gear of the planetary. The sun of the planetary is driven by the axial roller. Output speed and direction are determined from the relative rotations of the sun and ring gears.

3.1.5 Disc

The cone disc traction drive was invested by Joseph Beier of Austria, who was the chief engineer for the Brown Boveri Co. in Mannheim, Germany.

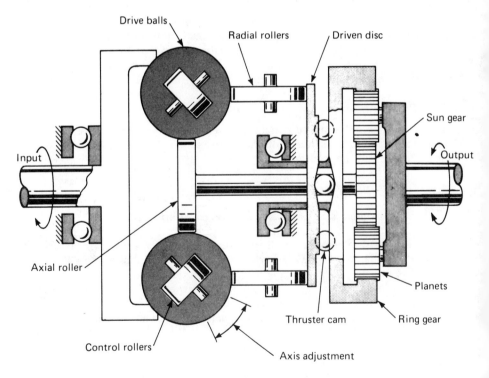

FIG. 3.44 Schematic of the Floyd free-ball drive. (Courtesy of Product Engineering, Morgan-Grampian Inc., New York.)

Types of Traction Drives 77

Sumitomo of Japan has been licensed since 1952 to produce the Beier variator, and in 1954, they started extensive field testing which lead to a more conservative design unit which is capable of being supplied in sizes to 200 hp. Two other manufactures of disc traction drives are Shimpo Industrial Co. of Japan and the Disco drive by Maschinenfabrik Hans Lenze of Germany. The Sumitomo drive closely follows the Beier principle. The Shimpo and Lenze units are planetary drives.

All three drives contain variable-speed ratio elements consisting of double-sided thin, flat cones that move in and out of mating discs, for which the contact pressure is provided by a torque-responsive cam loading mechanism.

In the Sumitomo drive, the cones are rotated by idler gears from the input shaft and mate with the output discs splined to the output shaft and the speed ratio is determined by the depth of engagement between the cones and discs. The contact areas are large, with the resulting contact pressure being lower than normally encountered with traction drives.

The Shimpo unit is a planetary drive with the movable discs mating with the outer stationary reaction discs and the inner discs, which are on the input shaft. The output shaft is attached to the carrier of the cones, which act as the planets. The speed ratio is determined by the depth of contact the cones make with the input and reaction discs.

The Lenze Disco drive works similarly to the Shimpo units.

Beier Variator Cone Disc

The Sumitomo Beier variator cone disc drive consists of alternating sets of cone driver discs and flanged driven discs.

Operating features

1. There is relatively low contact pressure since power is transmitted over many contact points.
2. Power is transmitted through the viscous friction of a film of oil that separates multiple thin metal discs.
3. Speeds are infinitely adjustable within the range; and settings remain constant, even over long periods of time.
4. The efficiency approaches 90%; high efficiencies are maintained even when combined with speed reducers.
5. Excellent shock load resistance results from the high multiplicity of contact points.
6. The sizes available are from 1/2 to 200 hp.
7. It is nearly a constant-horsepower drive.
8. The speed ratio can be changed only while in motion, not while at rest.

9. The drive must be mounted horizontally, but special units can be mounted vertically.
10. Shaft rotation of the output is opposite that of the input on the basic unit.
11. The input and output shafts are coaxial.
12. The speed ratio is 3.3:1 and 4:1.
13. The drive can be equipped with the Sumitomo Cyclo reducer to give very low output speeds.
14. It can be operated in either direction of rotation.

Principles of operation. The rotation of the input shaft (Fig. 3.45) is transmitted through the idler gear to the cone disc aligned on a multiple set of splined countershafts. These cone discs are driven between the flange discs connected to the output shaft. The motion of the cone discs is transmitted to the flange discs and then through the face cam to the output shaft.

The torque-sensitive face cam, complemented with a spring, automatically creates a disc contact pressure corresponding to the load torque, for the purpose of preventing excessive slippage due to load, and reducing pressure under light load. The multiple countershafts, on which the cone discs are mounted, are connected by the swiveling arms. Rotational movement of the swiveling arms around the idler gear forces the centers of the countershafts with the cone discs to move toward or away from the centers of the flange discs. A circular shifting ring connecting these swiveling arms, to the shifting screw, enables the movement of each swiveling arm in unison. When the point of contact of the flange discs is near the periphery of the cone discs, high output speed is attained, and when the point of contact of the flange discs is near the center of the cone discs, low output speed is attained.

Power transmission is conducted through the viscous drag or the tractive force of the oil film between the discs at the point of contact (Fig. 3.46). The cone discs are relatively thin, and the radius of curvature is great. This minimizes the contact pressure. Accordingly, at the points of contact, an ideal boundary lubrication nearing fluid film friction is obtained.

Since power transmission is evenly distributed over a number of contact points, the contact pressure is relatively low. For this reason, the Beier can withstand momentary shock or heavy overload and still not break the oil film, which would result in metallic contact. Slip rate at full load ranges from 4% at low speed down to 2% at high speed (Fig. 3.47). It has been theoretically and experimentally proved that in variable-speed drives utilizing viscous drag friction, excessive slip and poor speed regulation may result in unacceptably high power losses. The Beier variator is designed to minimize slip through main-

Types of Traction Drives

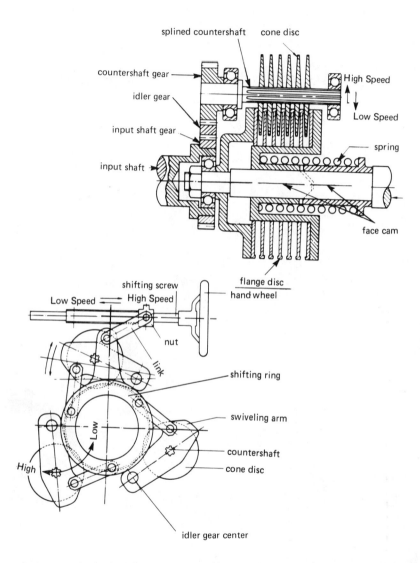

FIG. 3.45 Schematic of the Sumitomo Beier disc variator. (Courtesy of Sumitomo Machinery Corporation of America, Teterboro, N.J.)

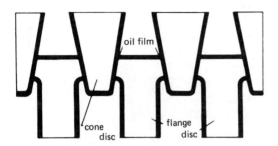

FIG. 3.46 Enlargement showing the interface between the cones, oil, and discs of the Beier variator. (Courtesy of Sumitomo Machinery Corporation of America, Teterboro, N.J.)

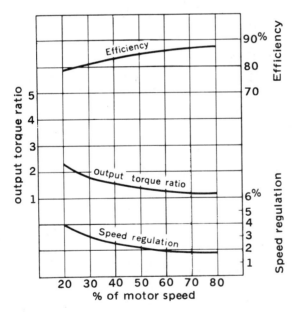

FIG. 3.47 Performance curves of the Sumitomo Beier variator. (Courtesy of Sumitomo Machinery Corporation of America, Teterboro, N.J.)

taining proper contact pressure under varying load conditions. Efficiency ranges from 79% at low speed to 87% at high speed (Fig. 3.47).

Planetary Cone Disc (Lenze)

The Lenze Disco drive consists of a planetary-type arrangement of driver discs, planet cone discs, and fixed reaction discs with the output through the planet disc carrier.

Operating features

1. The drive is a planetary disc arrangement.
2. The drive is available in sizes from 1/3 to 25 hp.
3. It can be mounted horizontally or vertically down.
4. The speed ratio is 6:1 (i.e., 9:1 to 1.5:1).
5. The input end output shafts are coaxial.
6. The drive is adjustable in output from near zero to +2100 rpm.
7. It can be adjusted in motion or to a lower speed at rest.
8. Efficiency is from 75% at low speed to 84% at high speed.
9. Flange- and foot-mounted styles are available.
10. It is available with reducers.
11. The slip varies from 3 to 7% depending on the load and speed.
12. The reset accuracy is within 1%.
13. It can be operated in either direction of rotation.
14. The output shaft rotates in opposite direction to the input shaft.

Principles of operation. Figure 3.48 shows the power flow from the input shaft through the sun assembly, to the planet set and planet carrier to the output shaft. Figure 3.49 shows how the unit's efficiency increases with an increase in output speed for a constant input speed.

The Disco planetary variable-speed drive is a planetary drive (Fig. 3.48). The input shaft (1) or the motor shaft (Fig. 3.50) drives the left inner sun (2) at a constant speed. The shaft drives the inner sun by a key. The left inner sun (2) and the right inner sun (3) are held together by the cup spring cluster (4). The planets (5) are are retained between the sun halves (2 and 3). The planets can number between 3 and 8 depending on the size of the unit. The axles of the conical planets (5) are supported in planet bearings (6). The planets (5) are also retained between the stationary outer ring (7) and the movable outer ring (8). The stationary outer ring (7) is kept in place by the drive housing. The movable outer ring (8) is adjustable both radially and axially. The input shaft (1) turns the inner sun halves (2 and 3), which in turn rolls the planets (5) between the outer rings (7 and 8). The planets (5) rotate about their own axis in the planet bearings. They also rotate the planet carrier (11), which

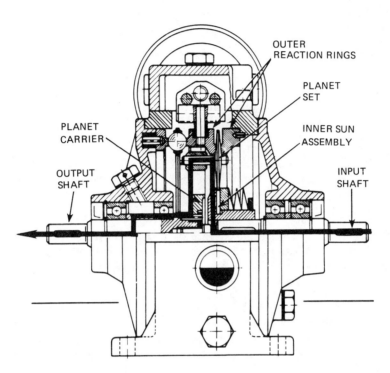

FIG. 3.48 Schematic of the Lenze Disco disc variator showing the power flow. (Courtesy of Simplana Corporation, Ann Arbor, Mich.)

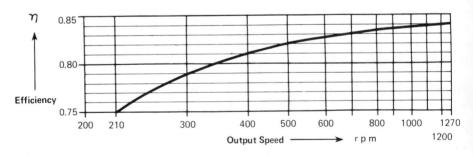

FIG. 3.49 Efficiency curve at 100% load for the Lenze Disco drive. (Courtesy of Simplana Corporation, Ann Arbor, Mich.)

Types of Traction Drives 83

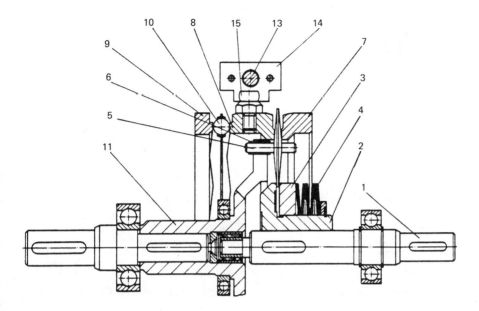

FIG. 3.50 Schematic of the Disco drive components. (Courtesy of Simplana Corporation, Ann Arbor, Mich.)

houses the planet bearings (6). The planet carrier (11) then turns the output shaft (16) by means of a key. Variable speed is obtained by varying the gap (e) (Fig. 3.51) between the outer rings (7 and 8). The planets will run on a small or large running diameter (d_1 or d_2), depending on the gap (e). The gap (e) is varied by rotating the movable outer ring (8) with the control shaft (13) (Fig. 3.50), guide (14), and speed control stud (15). The control stud (15) threads into the movable outer ring (8). The movable outer ring (8) has a cam profile, thus converting radial movement into axial movement. A ball cage (10) is located between the movable outer ring (8) and the cam ring (9) for ease of adjustment. An increase in the gap (e) (Fig. 3.51) between the rings allows the planets (5), via cup spring pressure and centrifugal force, to run in a large running diameter (d_1). A reduction of the gap (e) forces the planets (5) to a smaller running diameter (d_2). At the smallest running diameter (d_2), the planets (5) run at the fastest speed, which in turn causes the output shaft (16) to turn at its fastest speed. The largest running diameter (d_1) causes the output shaft to turn at its lowest speed. The ratio between the input speed and the maximum output speed is 1.5, and between the input speed and the minimum output speed is 9. Therefore, the overall variable speed range is 6:1.

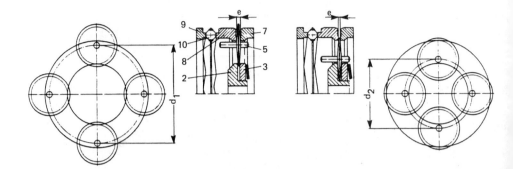

FIG. 3.51 Schematic of the Lenze Disco planetary disc drive parts relationship for (left) low and (right) high speeds. (Courtesy of Simplana Corporation, Ann Arbor, Mich.)

Planetary Cone Disc (Shimpo)

The Shimpo planetary disc drive is very similar to the Disco unit by Lenze.

Operating features

1. The drive is a constant-horsepower device.
2. The speed ratio is 5:1.
3. Infinitely variable speed adjustment from 1 to 1/5 of input speed.
4. The torque capacity is maintained substantially regardless of the load variations over the entire speed range due to the compound friction system.
5. The drive is available in sizes from 3 to 20 hp.
6. The input and output shafts are coaxial.
7. The drive can be operated in either direction of rotation.
8. The output shaft rotates in the opposite direction from the input shaft.
9. It can be flange or foot mounted.
10. The drive can be mounted horizontally or vertically down.
11. The speed can be changed only while in motion, not at rest.

Principles of operation. Power is transmitted from the input shaft (Fig. 3.52) to input rings in contact with cones pressed by rotatable output rings. As they are pressed between input rings and output rings, the cones rotate on their axles and transmit the rotation of the input rings to the output rings. The rotation of the output rings transmits through the ring holder and carrier the output.

Types of Traction Drives   85

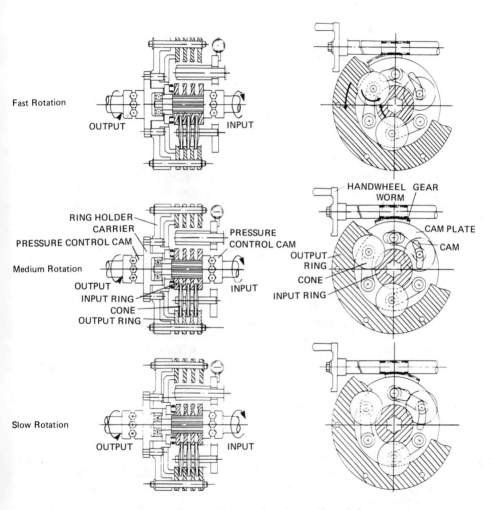

FIG. 3.52 Schematic of the Shimpo planetary disc drive component position for high, medium, and slow speeds. (Courtesy of Shimpo Industrial Co., Ltd., Kyoto, Japan.)

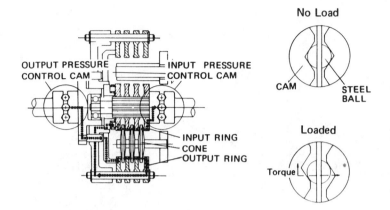

FIG. 3.53 Schematic of the power flow and torque-loading mechanism of the Shimpo disc drive. (Courtesy of Shimpo Industrial Co., Ltd., Kyoto, Japan.)

Output speed is varied by moving the cones along the cams and changing the depth to which the cones are inserted between the input rings, thereby altering the effective diameter of the cones.

The pressure control mechanism (Fig. 3.53) is essential for efficient transmission of power and for protection of overload. The pressure control mechanism consists of double cams and steel balls, which are located on both input and output shafts. By following the dotted arrows, the operation maintains proper pressure from the rings to the cones on the input side, and through the carrier, thrust disc, and rings to the cones on the output side. The ring holder and carrier are pressed by the action of a "wedge." This applies pressure to the friction surface of the transmission. The amount of pressure automatically adjusts according to the fluctuation in the load.

3.1.6 Toroidal

The toroidal traction drive has received much attention over the years and has been in existence since 1877, when Charles Hunt received a patent for a friction drive. In 1922, Richard Erban applied for a patent for a traction drive, and in 1925, Frank Hayes filed a patent for a dual toroidal traction drive.

Numerous other patents have been issued for improvements to the basic concept, including several to Charles Kraus of Excelermatic for use as a replacement for the automotive automatic transmission. Several industrial toroidal drives include those of Transitorq by General Motors, by Excelermatic in Austin, Texas; the Sadivar by

Types of Traction Drives

David Brown Sadi in Belgium; by Arter Regelgetriebe in Switzerland; and by Metron Instruments Inc. in Denver, Colorado, which was used for instrument drives.

Toroidal drives consist of input and output toroidal discs with tilting rollers inside the toroidal cavity. The speed ratio is controlled by steering the rollers on the discs. Toroidal drives have been developed up to 220 hp, with an overall efficiency exceeding 90%.

Single-Cavity Two-Roller Toroidal

The Arter toroidal traction drive's power transmission is obtained by the pure rolling movement between components.

Operating features

1. The unit can be driven in both directions.
2. Motors can be mounted directly.
3. The efficiency is up to 95%, depending on the speed ratio setting.
4. Output shaft is coaxial with the input shaft.
5. Either shaft can be used for input.
6. It can be mounted horizontally or vertically, but must be so specified; the horizontal unit cannot be used vertically.
7. Reduction gear ratios are available.
8. The drive is available in sizes from 0.20 to 15 hp.
9. The drive can handle a 2.5-fold starting overload capacity.
10. Speed ratios are from 7:1 to 10:1, depending on the size selected.
11. Speed range is from 0 to 2200 rpm with a planetary gear arrangement on the output.
12. It is a constant-horsepower drive.
13. The drive can be flange or foot mounted.
14. The output shaft rotates in the opposite direction from the input shaft.
15. The speed ratio can be adjusted while at rest or when in motion.

Principles of operation. The rolling members are shaped so that tangent lines drawn from their points of contact intersect with each other very close to the main axis of rotation for all speed ratios. The size of the power transmitting members can be calculated with the equations of Hertz.

Figure 3.54 shows a section through the variable-speed drive. The power enters through shaft (1) and cone (3). The cone is in contact with the two pivot wheels (20), which turn the drive cone (4) and output shaft (2). For changing the speed ratio, the pivot wheels

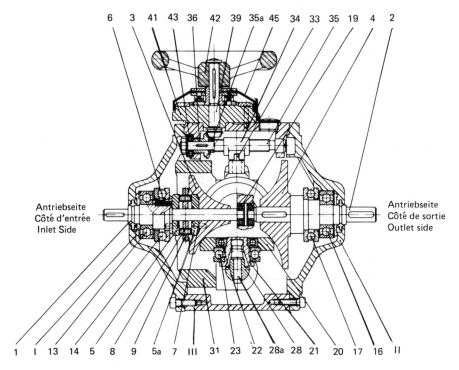

FIG. 3.54 Schematic of the Arter toroidal drive. (Courtesy of Arter Regelgetriebe Ltd., Männedorf, Switzerland.)

swivel around the vertical axes of the levers (28). The lever itself receives its motion through the handwheel (42) and feedscrew (35). The pivoting levers (28) are mounted in a support (31) which is free to adjust itself, allowing the forces acting on the pivoting wheels to be equalized. In their axial direction, the pivot wheels (20) are resting against the curved faces of the pivots (28a), permitting adjustment between the two cones (3 and 4), equalizing the two forces acting on the pivoting wheel from this direction also. The driven cone (4) is mounted on the output shaft with a drive key. The input cone (3) is driven through a loading mechanism consisting of the two pressure plates (5 and 5a) as well as the pressure rollers (8). The screw faces on the pressure plates provide torque-responsive pressure to cones (3) and (4) and wheels (20). The springs (6) supply only the pressure needed when the drive is running idle.

For a drive equipped with the planetary output gear set, the input shaft extends to drive a sun gear and the output cone drives

Types of Traction Drives

the ring gear, and the final output is from the planets to the planet carrier to the output shaft.

Figure 3.55 shows the positioning of the actual drive components and Fig. 3.56 shows the pitch cone angle position for a unit with equal-size driver and driven cones, and one with a smaller driver cone than driven cone.

Toroidal with Planetary

The Metron instrument drives were offered as straight toroidal drives or equipped with a planetary gear set for zero speed at midrange, then reversing to operate low-power instrument devices. The Metron drives are no longer available. The principle of operation was the same as that of the Arter units for the straight toroidal drive and the unit with the planetary output. Maximum speed was 4000 rpm for the planetary drive and 10,000 rpm for the straight drive.

Single-Cavity Three-Roller Toroidal

The David Brown Sadi Sadivar drive (Fig. 3.57) incorporates three sets of rollers between the input and output toroidal cone faces. One roller is tilted to control the speed ratio while the other two rollers assume symmetrical positions (dynamic balance) by means of a floating

FIG. 3.55 Photograph showing the relationship of the actual parts of an Arter drive. (Courtesy of Arter Regelgetriebe Ltd., Männedorf, Switzerland.)

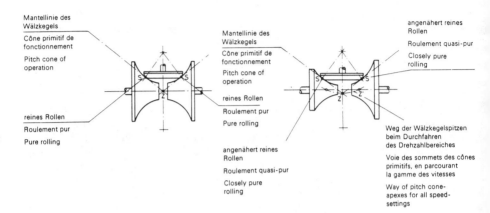

FIG. 3.56 Pitch cone angle for equal and unequal cones of an Arter toroidal drive. (Courtesy of Arter Regelgetriebe Ltd., Männedorf, Switzerland.)

FIG. 3.57 Cutaway of the David Brown Sadi Sadivar toroidal drive. (Courtesy of David Brown Gear Industries, Inc., Ontario, Canada.)

Types of Traction Drives

FIG. 3.58 Close-up of the steering mechanism of the Sadivar drive. (Courtesy of David Brown Gear Industries, Inc., Ontario, Canada.)

assembly with a slide bar and a self-centering knuckle assembly (Fig. 3.58).

Operating features

1. The output shaft is coaxial with the input shaft.
2. The speed ratio is 7:1 with 1:1 at the midrange.
3. The maximum efficiency is 80% at a 1:1 speed ratio.
4. The speed range is 540 to 3780 rpm at 1750-rpm input.
5. Four sizes are available from 0.33 to 5.50 hp.
6. The drive is a constant-horsepower device.
7. The unit is available foot or flange mounted.
8. The drive may be run in either direction of rotation.

9. Lubrication is by oil bath for horizontal operation and oil pump for vertical operation.
10. The maximum slip between discs and rollers under all operating conditions is 3%.

Single-Cavity and Dual-Cavity Toroidal

The Excelermatic industrial toroidal traction drive speed ratio is controlled by tilting the multiple rollers inside the toroidal cavity formed by the input and output toroidal cone discs. The rollers are steered hydraulically to obtain the desired speed ratios (see Figs. 3.59 and 3.60).

Operating features

1. The speed range is 9:1 (i.e., 1:3 to 3:1 of the input speed).
2. The rollers are hydraulically balanced to carry equal load under all operating conditions.
3. A torque loading cam is designed to protect the unit from shock of up to two to three times the normal load.
4. The drive can be assembled with either negligible or significant speed droop from no load to full load.
5. The unit is rated at 220 hp up to 250 hp at 8000 rpm.
6. The efficiency peaks at 95 to 96% and exceeds 90% efficiency with a 10 to 15% overload above the rated load.

Principles of operation. The power is transmitted from the input toroidal disc to the output toroidal disc by the traction forces between the discs and the rollers. The angle of the rollers determines the rolling radii of the input and output discs and the speed ratio between the input and output shafts.

The normal forces between the discs and the rollers are varied with varying torque and ratio conditions. This is accomplished by means of an initial preload force from springs to prevent any initial slip. Upon rotation and torque application, the shaft rotates a load cam which pushes against one of the discs, increasing the normal force as a function of torque.

The rollers are supported through bearings by flexible supports (Fig. 3.59). The flexure-type structure is required to enable deflection of the roller assembly perpendicular to the axial direction of the discs to permit proper steering of the rollers required for changing the roller angle and ratio.

The rollers are steered and held in position by means of hydraulic control pistons which are connected in parallel through a hydraulic circuit. This hydraulic force balances the tangential forces on the rollers. When a new ratio position is desired, the hydraulic pressure

Types of Traction Drives

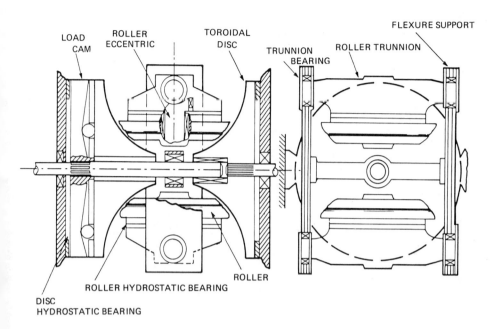

FIG. 3.59 Schematic of the Excelermatic single-cavity toroidal drive. (Courtesy of Excelermatic, Inc., Austin, Tex.)

is changed in the control cylinders, causing the rollers to move from the tangent position of roll and develop a vectorial steering error. This causes the rollers to steer to a new position until the forces are balanced and the rollers again return to the tangent point on the roll. The parallel hydraulic connections between the roller control cylinders enable all rollers to share the same loads.

Each power roller is supported on its trunion through an eccentric which allows the roller to move relative to the trunion. This permits the roller to track properly for variations in the toroidal cavity position which result from load changes. The toroidal discs shift due to the load cam action and the deflection of the parts themselves. The eccentric can be assembled to increase or decrease the load droop. The offset roller position geometry enables the control of spin between the rollers and discs to acceptable limits of less than 4%.

Single-cavity toroidal drives develop high normal forces coincident with the axis of the toroidal drives. These forces must be supported through thrust bearings to the housing. Off-center roller drives develop thrust forces on the roller supports which are not generated in on center roller drives. The forces are carried in tension in the flexure supports and are not transmitted to the housing. Although

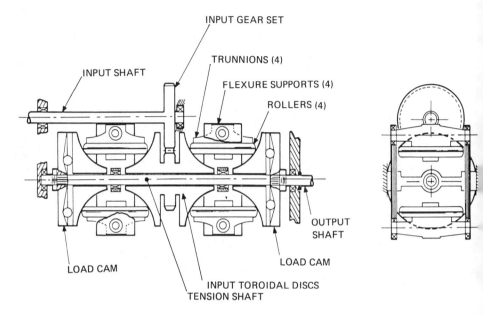

FIG. 3.60 Schematic of the Excelermatic dual-cavity toroidal drive. (Courtesy of Excelermatic, Inc., Austin, Tex.)

the on-center drives do not develop these thrust forces, they induce spin, which destroys the contact surfaces.

Another approach to handling these high normal or axial forces is by using a double-cavity drive (Fig. 3.60). The axial forces are balanced between the outboard discs and the tension shaft connecting the discs. The double-cavity drive is about 5% more efficient than the single-cavity drive. The reason for the increased efficiency is the elimination of the large hydrostatic thrust bearings used in the single-cavity drive that cause lubrication losses.

3.1.7 Planetary

The planetary traction drive is broken down into two categories, fixed ratio and adjustable ratio. The fixed-ratio variations discussed are those manufactured by Wedgtrac Corp. of Yorkville, Illinois, Nastec Inc. of Cleveland, Ohio; and Paxton Products of Santa Monica, California.

The adjustable-ratio traction drive examined will be the Chery drive developed by Walter Chery of Meadville, Pennsylvania, and put into production by Fafnir Bearing of New Britain, Connecticut. Drives

Types of Traction Drives 95

such as the Kopp roller variator and several of the Graham and Shimpo ring-cone variators, and the Disco and Shimpo disc variators, could also have listed as adjustable planetary traction drives.

Despite the relatively high torque capacity of fixed-ratio planetary traction drives, they have had limited commercial use. They have the ability to provide smooth, quiet power transfer at extremely high speeds and with good efficiency. Many earlier units were large and heavy compared to gear drives due to the poorer quality of steel and nonavailability of a suitable traction fluid. Another factor was the problem of slip and load-sharing roller elements.

Planetary Constant-Ratio Wedge Roller

The Wedgtrac planetary drive is a single-row planetary roller traction drive (Fig. 3.61) limited to one direction of rotation. The torque is transmitted by a wedging action between the input roll, the wedge roll, and the output ring.

Operating features

1. Roller contact supports the input and output shafts.
2. Normal full load causes a "creep" of approximately 0.5%. Output rpm equals 0.995 times input rpm divided by ratio of ring inside diameter to sun roller diameter.
3. The design is for unidirectional rotation. Note the arrows The drive overruns in reverse. Location of the wedge and reaction rolls are interchanged for opposite direction of rotation.
4. The offset of the input roll provides a wedging action which increases proportionally to the torque being transmitted. The wedge roll is spring loaded for initial contact.
5. Under torque overload, the wedge roll moves deeper into the drive pocket and the bore of the roll bottoms out against stationary stud. This causes drive contact between the sun roller and the three mating idler rolls to break loose and slip, thereby limiting torque. The value of the torque limit is controlled by dimensions and elasticity of parts.
6. The unit will tolerate several seconds of slip without damage.
7. The drives are custom-engineered special products rather than stock or off-the-shelf products.
8. The drive has the compact envelope of a planetary gear arrangement, but without the worry of equal load sharing between the planets.
9. There is a smooth continuous transfer of the tangential traction forces between ground rollers, with all radial contact forces being internally balanced.

10. The output shaft rotates in the opposite direction from the input shaft.
11. The efficiency is in excess of 90% for most speeds and ratios.
12. High-speed operation of up to 70,000 rpm is possible.
13. The speed ratio of single-stage reducers range from 3:1 to 20:1 and can operate as either reducers or increasers. Typical speed ratios are 3:1 to 8:1.
14. The input and output shafts are offset but parallel.

Principles of operation. The traction is established by transmitting tangential forces between two rotating bodies while engaged in rolling contact. Wedgtrac drives cannot experience gross slip. The only relative velocity evident in the contact zone between rollers is slight creep. Normal compressive forces at the contact zone must increase in proportion to tangential load demand if gross slip is to be avoided. Wedgtrac accomplishes this by the "offset" position of the input (sun) roller (Fig. 3.61). The arrangement of Wedgtrac rolling elements (Figs. 3.61 and 3.62) resembles a simple planetary drive with a sun input, ring output, and a fixed carrier. For the direction of rotation shown, the upper left-hand roller floats freely in dynamic equilibrium and transmits power by a wedging action between the sun and internal ring rollers. This wedging action preloads the entire system in proportion to transmitted load. The two remaining planet

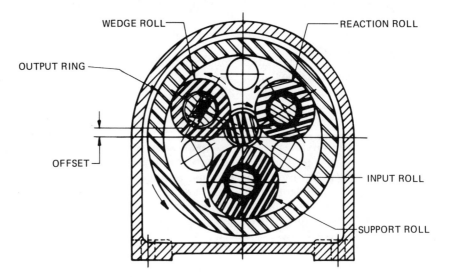

FIG. 3.61 Schematic of the Wedgtrac wedge roller fixed-ratio drive. (Courtesy of Wedgtrac Corporation, Yorkville, Ill.)

Types of Traction Drives　　　　　　　　　　　　　　　　　　　　97

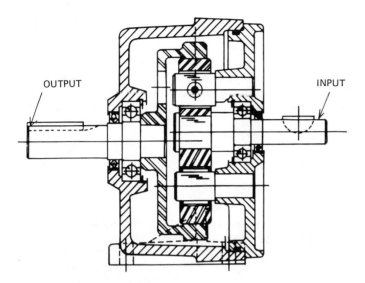

FIG. 3.62 Side view of the Wedgtrac fixed-ratio drive. (Courtesy of Wedgtrac Corporation, Yorkville, Ill.)

rollers become reaction members and operate on relatively fixed bearing supports. Both the sun and ring members have sufficient degrees of freedom to ensure automatic response to the preload imposed by the wedging roller. Traction Propulsion, Inc., developed a wedge roller drive which paved the way for similar wedge-type units.

Planetary Constant-Ratio Roller

The Nastec planetary roller traction drive (Fig. 3.63), is a multirow multiroller drive invented by Algirdas Nasvytis in the mid-1960s, with further development by NASA's Lewis Research Center under the leadership of Stuart Loewenthal. The Nasvytrac unit has been used on exotic drive systems requiring high speeds, high ratios, high torques, low noise, low weight, and small size.

Operating features

1. Ratios up to 150:1 are available (e.g., 3.25:1, 15:1, and 48:1).
2. Units have operated to 480,000 rpm.
3. The drive acts as a bearing, thereby reducing the number of bearings that would normally be required.
4. The drive has a compact planetary arrangement.

5. The stepped planet rollers provide a large speed ratio in a single stage.
6. The multiplanet roller contacts share the contact loads.
7. The input and output shafts are coaxial.
8. The internal loads are balanced through floating rollers.
9. The unit has an automatic torque-loading mechanism.
10. The roller size, number, and position are optimized for required power, speed, ratio, and life.
11. The drive can be used as a speed increaser or speed reducer.
12. It is available in sizes from 10 to 500 hp. A 3000-hp unit is under development for helicopter use.
13. The efficiency is over 95%.

Principles of operation. The basic Nasvytrac multiroller traction drive in Fig. 3.63 consists of two rows of five stepped planet rollers which are contained between the concentric high-speed sun and low-speed ring rollers. The sun roller drives the larger diameter of the first-row planet rollers. Their smaller inner diameters drive the second-row planet rollers, and these in turn rotate the ring rollers. The planet rollers do not orbit but are grounded to the case through reaction bearings contained only in the second or outer row of planets. This is a favorable position for the reaction bearings since the reaction forces and operating speeds are relatively low.

The sun roller and the first row of planets float freely, relying on contact with adjacent rollers for location. Because of this self-supporting roller approach, the number of total drive bearings are greatly reduced and the need for separate high-speed shaft support bearings has been eliminated. In addition, both rows of planets are in three-point contact, with adjacent rollers promoting a nearly ideal internal force balance. In the event of an imbalance in roller loading, the first and second rows of planets (supported by large clearance bearings) will shift under load until the force balance was reestablished. Consequently, slight mismatches in roller dimensions, housing distortions, underload, or thermal distortions merely cause a slight change in roller orientation without affecting performance. Because of this roller-cluster flexibility, the manufacturing tolerances set for roller dimensions can be rather generous relative to standards set for mass-produced bearing rollers.

The number of planet roller rows, the number of planet rollers in each row, and the relative diameter ratios at each contact are variables to be optimized according to the overall speed ratio and the uniformity of contact forces. In general, drives with two planet rows are suitable for speed ratios to about 35, and drives with three planet rows are suitable for ratios to about 150.

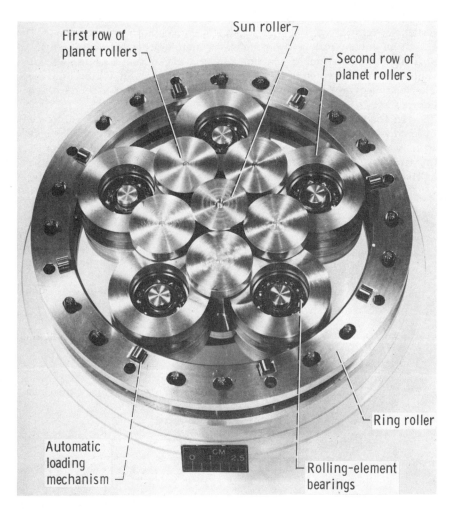

FIG. 3.63 View of the Nasvytrac constant-ratio roller drive showing the roller arrangement. (Courtesy of NASA Lewis Research Center, Cleveland, Ohio.)

There are two types of roller-cluster loading mechanisms: either in the ring assembly of the ring-loader drive or in the sun roller of the sun-loader drive. Both of these mechanisms automatically adjust the normal contact load between the rollers in proportion to the transmitted torque. These mechanisms operate above a preselected minimum preload setting.

The automatic loading mechanism ensures that there is always sufficient normal load to prevent slip under the most adverse operating conditions without needlessly overloading the contact under light loads. The load mechanism improves part-load efficiency and extends drive service life.

In the case of the ring-loader drive (Fig. 3.64) the loading mechanism consists of eight small rollers contained in wedge-shaped pockets, equally spaced circumferentially between the ring rollers and the backing rings. The inside diameters of the ring rollers and outside diameter of the second-row planets had slightly tapered (5.7°) contact surfaces.

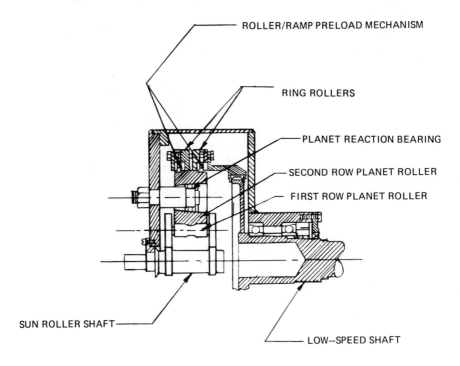

FIG. 3.64 Schematic of the power flow and components of the Nasvytrac constant-ratio roller drive. (Courtesy of Nastec Inc., Cleveland, Ohio.)

These centrally raised areas of the second-row planet rollers serve a second purpose of locating the first-row planetary rollers axially by engaging the depressed centers of these otherwise free-floating members. When torque is applied, the ring rollers will either circumferentially advance or retreat relative to the backing rings. This will cause the loading rollers to move up the ramped pocket, squeezing the ring rollers together axially and, in turn, radially loading the roller cluster through the tapered contact. The amount of normal force imposed on the traction drive contacts for a given torque can be varied by simply changing the slope of the wedge-shaped pockets.

Sun roller loading. The drive equipped with the sun roller loading mechanism uses the same principle, but loads the drive radially outward through a two-piece sun roller. Packaging the loading mechanism into the sun roller simplifies the drive design and reduces the cluster weight.

Ring loading. Generally speaking, the ring loader drives have a slightly higher peak efficiency values than do the sun loader drives, 96% versus 93% for the speed reducer. Part of this difference is attributed to the slightly tighter conformity of the planet and ring roller contact surface of the sun loader drive. The tighter conformity causes some reduction in contact stress, but results in slightly higher spin losses and creep.

Planetary Constant-Ratio Ball

Paxton Products' constant-ratio planetary traction drive was developed for driving a quiet, high-speed compressor. In the Paxton unit (Fig. 3.65) the input shaft drives through the planetary drive balls, with the ball driver acting as the carrier to the output shaft, which acts as the sun and the inner-race portion of a bearing. The outer race acts as the fixed reaction member of the planetary. The high contact forces and the loading mechanism through the split outer race and Belleville springs prevents slippage within the operating range.

Planetary Variable Ratio

The Chery variable-ratio planetary drive was invented with several variations by Walter Chery of Meadville, Pennsylvania, with further development by Arthur D. Little, Inc., of Cambridge, Massachusetts. A version of this drive was built by Chery and developed further by Fafnir Bearing in New Britain, Connecticut, for a tractor transmission application.

The Chery drive is a planetary traction drive using smooth rolling elements which can be coupled with a regenerative planetary gear set

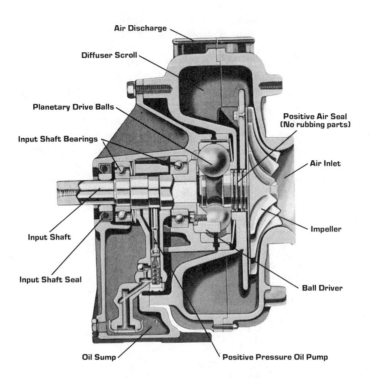

FIG. 3.65 Cutaway of the Paxton Products blower with a constant-ratio ball drive. (Courtesy of Paxton Products, Santa Monica, Calif.)

to yield speed variations from forward to zero to reverse. The unique feature of the Chery drive is its ability to automatically change its drive ratio in response to changes in load.

Operating features

1. The drive is a planetary self-aligning rolling element.
2. The speed ratio is automatically controlled by the output torque load.
3. Units have been built to 150 hp.
4. With a regenerative planetary, the drive output speed can range from forward to zero to reverse or from zero to forward.
5. The speed ratio is 4:1 on the straight unit and various combinations with planetary gearing.
6. It is a virtually constant-horsepower drive with full torque available at zero output speed.

Types of Traction Drives

7. The maximum operating speed is 5000 rpm.
8. The efficiency is 93% in high gear and 80% in low gear.
9. With low relative speed between the inner race, the planetary cage, and the output race of the planetary drive, the efficiency is nearly 97%
10. The output shaft is coaxial with the input shaft.
11. Depending on the variation of drive, the planet rollers can be individually spring loaded against races for uniform load distribution.
12. It is a true mechanical torque converter, since the torque multiplication ratio automatically varies steplessly in response to output load without any auxiliary control means, thus maintaining constant horsepower.
13. The same basic mechanism can be used as a stepless variable-speed drive by adding a ratio-adjusting device. The torque converter function can be retained in the variable-speed drive; that is, if a given output load is exceeded, the drive can automatically move to a greater ratio to avoid stalling or lugging the prime mover.
14. The pressure between the drive elements increases automatically in proportion to the output load; therefore, the drive resists slipping due to overloads or sudden load fluctuations. This loading is controlled by ball ramp cams and/or load springs.
15. In combination with planetary gears and overrunning clutches, the unit can provide a transmission which automatically steplessly reduces ratio during acceleration and then automatically locks up and rotates as a unit when a 1:1 ratio is reached, so that the friction drive elements are not working except during periods of acceleration. By adding a fluid coupling or hydraulic torque converter to the system, a slight relative rotation of the friction drive elements can be provided when running in the locked 1:1 condition.

Externally Adjusted Variable-Ratio Planetary

Principles of operation. Figure 3.66 will be considered as two parts, the planetary traction drive and the planetary regenerative gear set. The traction drive portion of the transmission is used to vary ratio, while the regenerative gears extend the ratio so that the transmission has variation to both forward and reverse ratios.

The input to the transmission drives both the two inner races of the traction drive and the sun gear of the regenerative gear set. This is a split-torque system with the traction drive ratios working against the fixed planetary gear set ratio. The ratio of the traction drive is varied by changing the axial spacing at the traction drive races. The axial location of the races determines the rolling radii of

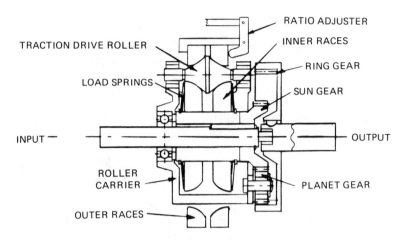

FIG. 3.66 Schematic of the Fafnir Bearing variable-ratio drive tractor transmission. (Courtesy of Fafnir Bearing, Division of Textron, New Britain, Conn.)

the inner and outer contacts, which establishes the traction drive ratio. The output of the traction drive is the roller carrier, and it is also the planet carrier for the regenerative gear set. The ratio and direction of rotation of the output ring gear are a result of the constant input speed of the sun gear and the variable speed of the planet carrier. Forward ratios are obtained when the traction drive overdrives the regenerative planets. Reverse ratios are obtained when the traction drive retards the regenerative planets. In neutral or zero output speed, the regenerative planets roll around the output ring gear.

Torque is transmitted through the smooth rolling element surfaces of the Chery drive by clamping the surfaces together and utilizing the traction properties of the contacting surfaces. One of the major advantages of this planetary-type traction drive is that the drive clamping loads are self-contained in the drive and support bearings are not required.

Belleville load springs are used to clamp the inner races together to apply the necessary load pressure. A balance spring bears against the outer races and balances out the reaction force of the Belleville springs.

To eliminate improper speed adjustment for the load, an automatic downshift mechanism can be added which overrides the manual control and sets the drive ratio according to the torque transmitted. Under extreme load, the drive will shift to zero output speed to protect the system from overload.

Internally Self-Adjusting Variable-Ratio Planetary

Figure 3.67 shows another basic Chery. In Fig. 3.67A, the unit is operating at minimum reduction ratio, and in Fig. 3.67B at maximum reduction ratio. The drive consists of multiple planetary rollers carried in a cage, rolling between a pair of inner sun wheels and a pair of outer races. The radius of curvature of the operating surfaces of the sun wheels and outer races is slightly greater than the radius of curvature of the rollers. The sun wheels are keyed to the input shaft. The rollers are preloaded axially outward against the sun wheels and outer races by Belleville washer springs. The outer races are free to rotate in the housing but are restrained by cams fixed in the housing, and balls, as shown in section B-B of Fig. 3.67. The planetary cage is attached to the output shaft.

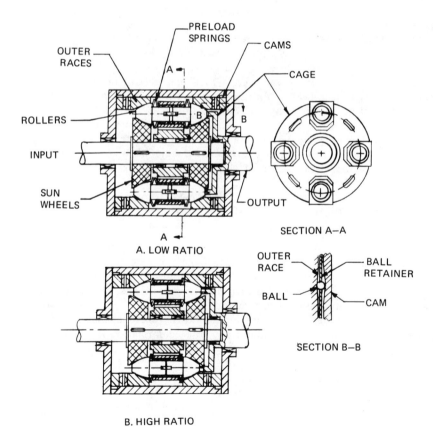

FIG. 3.67 Schematic of the self-adjusting Chery drive at low and high ratio settings. (Courtesy of Walter Chery, Meadville, Pa.)

In Fig. 3.67A, the unit is shown running at maximum output speed. If the output load is now increased, there will be a resulting proportional increase in the tangential force between the rollers and the outer races. This tends to rotate the outer races slightly in the housing, and this motion in turn causes the cams, acting through the balls, to move the races axially inward toward each other. This forces the rollers axially inward, compressing their preload springs, thus increasing the pressure between the friction elements. The roller assemblies also move radially inward, sliding in slots in the cage. As a result of the change in position of the races and rollers, contact between the rollers and races moves toward the small end of the rollers, and contact between the rollers and the sun wheels moves toward the large end of the rollers. This increases the reduction ratio of the drive, and output speed is reduced proportionally to the increase in output torque. The drive has automatically moved to the ratio required to carry the output load, and at this position, there is a balance between the output load and the preload spring forces.

To convert the basic mechanism to a steplessly variable speed drive, a means of positively adjusting the angular position of the outer races within the housing is added. This can be in the form of a nut and lead screw or a lever, operated manually or by powered operators such as electric motor or hydraulic or pneumatic cylinder. As the outer races are rotated in the housing, the cams force them axially closer together or allow the roller preload springs to force them farther apart, thus changing the speed ratio as described in the preceding paragraph. In a modification of the basic mechanism, hydraulic pressure can be used directly to vary the ratio by pushing the outer races axially closer together or allowing the preload springs to push them farther apart.

3.1.8 Spool

The spool traction drive got its name from the fact that the transmitting member between the input and output members looks like a spool. It is a low-power offshoot of the toroidal drive. It has a limited speed ratio, but operates at a high efficiency.

The spool units were developed by Traction Propulsion, Inc. of Austin, Texas, which is a subsidiary of Excelermatic, Inc., of Austin, Texas.

Operating features

1. The speed ratio is 2.5:1 to 1:2.5 input speed for an overall ratio of 6.25:1.
2. The output shaft rotates in the same direction as the input shaft.

Types of Traction Drives

3. The drive can be operated in either direction of rotation.
4. There is a locking-cam device to limit the effects of shock loading.
5. The input and output shafts are coaxial.
6. The horsepower ratings are from 0.20 to 15 hp for industrial use and to 50 hp for automotive application.
7. Input speeds of up to 10,000 rpm have been tested.
8. Efficiency is 93% at a ratio of 1:1 and drops to 85% at each extreme ratio. At maximum loads, the efficiency is lower at each ratio extreme.

Principles of operation. The spool traction drive operates similarly to the toroidal drives. The units (Fig. 3.68) consist of an input and an output cupped, O-shaped disc, with a movable spool between them. The spool's axis of rotation is offset from that of the discs.

The initial normal forces between the spool and cups are obtained by the Belleville springs which preload these members. As the torque is applied to the input shaft, the load cam pushes against the rear of one of the cup members, increasing the normal forces as the torque increases. The use of the load cam enables the drive to operate with

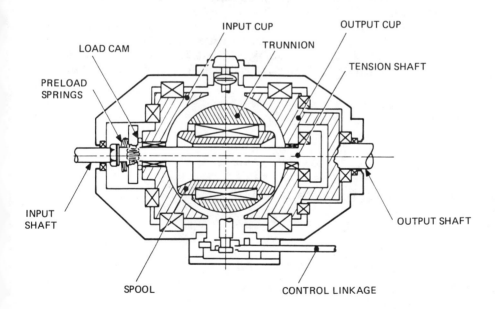

FIG. 3.68 Schematic of the Traction Propulsion, Inc., spool drive. (Courtesy of Traction Propulsion, Inc., Subsidiary of Excelermatic, Inc., Austin, Tex.)

the minimum normal loads sufficient to prevent slip under varying amounts of torque and shock loads. This results in increased life of the traction elements over preloading to a high normal force level that is always sufficient to carry the maximum torque expected. The spool is steered so as to produce different angles between the driver and driven cups to obtain the desired ratios between the input and output speeds.

The center shaft is a tension shaft which balances the loads between the load cam and the output cup, and hence no axial forces are transmitted to the housing when the drive is in a 1:1 ratio. However, when the drive is either in reduction or speedup, the angle of the spool changes to obtain the proper ratio (Fig. 3.69), thereby developing a component of force normal to the cup axis. This component is supported through the cup roller bearing to the housing. Since the contact forces occur only on one side of the shaft axis, a couple is formed which is also resisted through these bearings by the housing. This loading asymmetry, being a function of ratio position, accounts for the typical "bell"-shaped efficiency curves. The efficiency ranges from 85% at each extreme ratio to 93% at 1:1. The losses are due primarily to the loading on the bearings caused by the ratio and magnitude of the torque, thereby creating a family of efficiency curves as a function of torque and ratio.

The spool is steered to different angles by means of an eccentric to change the input-to-output speed ratio. This is accomplished by either a mechanical linkage or by a hydraulic piston which receives the command signal from the control system.

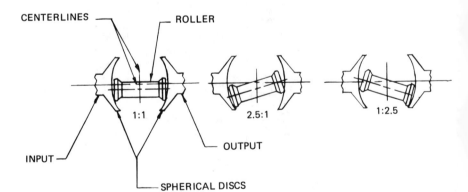

FIG. 3.69 Position of the spool with the input and output discs for various speed ratios. (Courtesy of Traction Propulsion, Inc., Subsidiary of Excelermatic, Inc., Austin, Tex.)

The steering eccentric also supports the tangential force couple generated on the spool and trunion as power is transmitted through the drive. Consequently, if a spring is used to support the steering eccentric, the drive becomes torque responsive.

3.1.9 Offset Sphere

The offset sphere traction drive was developed by the J. R. Young Corp. of Racine, Wisconsin. The offset sphere drive is highly efficient and the speed is changed by varying the position of the movable sphere carrier with the offset spherical input and output shafts.

Operating features

1. The output shaft is offset but parallel to the input shaft.
2. A loading cam is provided to maintain pressure.
3. The drive axes are supported by bearings.
4. The efficiency is 95%.
5. The overall speed ratio is 20:1.
6. The speed ratio can be adjusted while the unit is at rest or while in motion.
7. There is little variation of speed under load.
8. The capacity is to 20 hp.
9. A version was built to go through zero speed.

Principles of operation. The offset sphere traction drive (Fig. 3.70) consists of two facing sphere segments, input and output, that rotate on parallel but slightly offset shafts. An adjustable carrier supports two similar spherical segments, one of which is in contact with the input segment and the other with the output segment. The speed ratio is changed by moving the carrier across the driver and driven axes. This changes the effective radius of the mating spherical segments. The drive axes are bearing supported. The contact forces between the mating surfaces is obtained and maintained by both springs and a loading cam. These drives are no longer available.

3.2 FRICTION DRIVES

Although not traction drives as considered by this chapter, two currently available friction drives will be discussed because they are similar to devices used early in the development of traction drives. These two drives each contain a driver and a driven element held in intimate contact with each other and transmit power by friction without the benefit of a traction fluid.

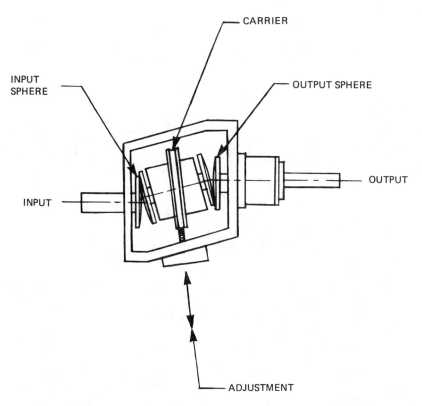

FIG. 3.70 Schematic of the J. R. Young offset sphere drive. (Courtesy of *Machine Design*, The Penton Publishing Company, Cleveland, Ohio.)

The two friction drives are the Varimot variable-speed drive manufactured by Eurodrive Inc. of Troy, Ohio, and the friction drive manufactured by McDonough Power Equipment of McDonough, Georgia.

Parallel Disc and Ring Friction

The Eurodrive Varimot friction drive transmits torque through contact between a constant-speed steel disc and a traction ring made of a synthetic material similar to a brake lining (Fig. 3.71). The torque is transmitted in a nonlubricated air-cooled environment.

Friction Drives

FIG. 3.71 Cutaway of the Eurodrive Varimot friction drive. (Courtesy of Eurodrive Inc., Troy, Ohio.)

Operating features

1. The sizes available are 1/4 to 15 hp.
2. The speed range is 5:1.
3. Output speeds to 4000 rpm are obtainable.
4. A rugged totally enclosed fan-cooled motor is mounted on an adjustable bracket.
5. The conical driven disc is mounted on the motor shaft.
6. The traction ring is steel with a self-cleaning friction face.
7. The screw-type control provides the means to vary the position of the motor-driven disc in relation to the traction ring.
8. Variable speed is obtained by changing the relative running diameter of the driven disc and traction ring by moving the motor-driven disc in a parallel path either closer to or farther away from the centerline of the output shaft.

9. A torque compensator incorporated into the output assembly matches the pressure between the driven disc and the traction ring to the torque required by the load, minimizing wear and providing protection against damaging shock loads.

Principles of operation. Torque is transmitted through contact between a constant-speed motor-driven disc and a variable-speed friction ring. The steel drive disc is mounted on a constant-speed induction motor rotor shaft (Fig. 3.72). The disc is slightly conical in shape, and the motor assembly is located slightly off the horizontal centerline. The conical disc and motor position therefore provides a linear contact with the driven friction ring throughout its speed range.

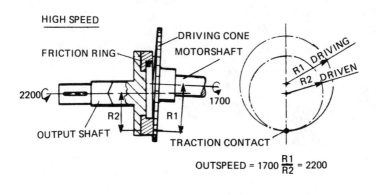

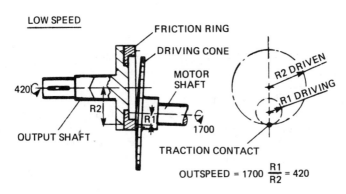

FIG. 3.72 Schematic of the speed change mechanism of the Eurodrive at high and low speeds. (Courtesy of Eurodrive Inc., Troy, Ohio.)

Friction Drives

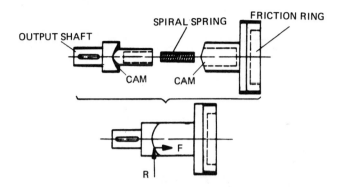

FIG. 3.73 Cam torque-loading assembly of the Eurodrive. The Cam action transforms radial forces (torque demands) R into axial forces F, increasing the traction contact. The constant spring force has only an auxiliary function. (Courtesy of Eurodrive Inc., Troy, Ohio.)

For speed variation, the constant-speed assembly is moved against the fixed-position driven friction ring assembly. This changes the ratio of centers of rotation R_1 and R_2, and therefore the speed. Since some slippage must occur between the driven ring and the driving cone, there will be a variation in speed with load.

In order to transmit the radial force over the area of friction contact (Fig. 3.73), a force perpendicular to the friction contact is needed. This force is produced by two means:

1. Spring loading.
2. A variable force, produced through an internal cam action which couples the output shaft to the traction ring assembly. This is a function of the rotary torque demand (the required load), which is the action, and an axial force, which is the reaction.

By means of this torque loading, the heat generation, bearing stresses, and wear are held to a minimum.

The cam action is designed such that slippage will occur at torque levels approximately 2.5 times the motor's rated load torque. The friction contact area is actually a rectangular shape. Due to the difference of the circumferential speeds on the inside and on the outside of that rectangular area, a rotary motion pattern is imposed on the whole area. This rotary motion is zero at speed ratio 1:1 (output speed equals motor speed), but increases with increasing

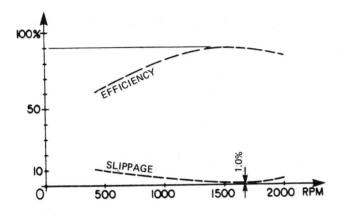

FIG. 3.74 Eurodrive efficiency and slippage curves. (Courtesy of Eurodrive Inc., Troy, Ohio.)

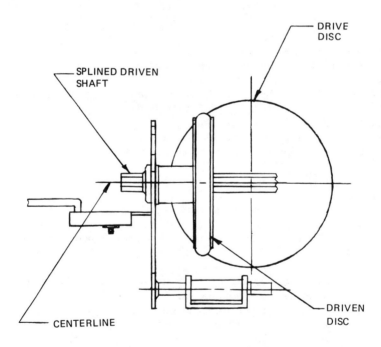

FIG. 3.75 Schematic of the McDonough power equipment friction drive used on Snapper riding mowers. (Courtesy of McDonough Power Equipment, McDonough, Ga.)

difference between R_1 and R_2, being at its maximum at minimum speed, making the drive less efficient at that position.

This rotary motion and the losses generated do not contribute to the torque supply and generate only heat. To take care of the greater amount of heat generated, especially at slow speed, the torque available at the shaft has to be reduced.

Care must be exercised in the use of the friction drives such as where high inertial loads must be started from rest or where acceleration of the unit takes longer than 3 sec, as these situations will cause excessive heating and damage the synthetic material of the friction ring.

Figure 3.74 shows the relationship of the percentage of slippage and of efficiency to output speed with a constant input speed of 1750 rpm.

Perpendicular Disc Friction

The friction drive produced by McDonough Power Equipment is used on the Snapper riding mowers and is similar to the type used on the early automobile friction drives and also on the railroad motor cars. It consists of a hardened-steel drive disc and a rubber-tired driven disc (Fig. 3.75). The driven disc is mounted perpendicular to the axis of the drive disc.

The speed ratio is changed by sliding the driven disc on a splined shaft across the face of the drive disc. The farther the driven disc is from the center of rotation of the drive disc, the greater the speed. When the driven disc is in the center of the drive disc, there is just spin with no output speed or torque transmitted. Moving the driven disc to one side of the center of the drive disc produces a desired forward motion, and moving to the other side of center gives the reverse motion.

REFERENCES

The references used in the preparation of this chapter on the generic types of traction drives are listed in Appendix A. These include the catalogs and the various papers that are available from the manufacturers of the drives or from those who have done extensive research on traction drives. These sources are listed according to the individual type of drive.

APPENDIX Traction Drive Characteristics

Type	Supplier	Horsepower rating	Speed range (rpm)	Speed change at load (%)	Efficiency (%)	Ratio
Kopp Ball	Winsmith	1/3-15	600-5400 @ 1800 input	0.1-4	75-90	(3:1-1:3) 9:1 total
	Kopp; Koppers	1/4-2	Varies with size, but 250-3000 @ 1500 input		82-90	(4:1-1:2) 8:1 total @ const. hp (6:1-1:2) 12:1 total @ const. torque
	Cleveland Gear	0.8-16	600-5400	0.1-4	75-90	(3:1-1:3) 9:1 total
			600-3200 @ 1800 input 0.75-1.25 times input			(3:1-1:2) 6:1 total Limited range
Roller	Kopp; Koppers	1-110	250-3000 @ 1800 input	1-4, 0.03% drift	90 @ 1:1 80 @ each extreme	(7.18:1-1:1.67) 12:1 total
Ball disc, multiball	P.I.V. Antrieb Werner Reimers	0.33-4	0-1.2 times input	1-3	Up to 85 85	0-1:1.2 (8.33:1-1:1.2) 10:1 total
	Graham	0.25; 0.75	0-1.5 times input	4	50	0-1:1.5
Roller disc Single roller	Parker/F.U. Unicum	0.5-4	340-2040 @ 1750 input 600-1800 @ 1750 input 720-1440 @ 1750 input	0.3	77-92	(3:1-1:1) 3:1 total, 6:1, 8.5:1, 10:1
Four rollers	Parker/F.U. Unicum	4-100		0.3	87-93	(2.4:1-1.22:1) 2:1 total, 3:1, 6:1, 8:1
Ring cone Planetary	Graham	1/12-5	Varies with size, but 0-550 @ 1750 input; a special reverses	0.1	85-90	0-3:1
Single cone	Graham; Shimpo	0.25-20	112-450 @ 1800 input		82	4:1 adjust with 4:1 built in
Dual cone	Shimpo	0.125-5	-360 to +360 @ 1800 input		60	10:1 forward 12:1 reverse
Parallel cone w/o planetary	Shimpo	1/16-3	600-2400 @ 1200 input			(2:1-1:2) 4:1 total
Ring with movable cone	Shimpo	1/16-2	150-1500 @ 1800 input			10:1
Variable cone with ring	Heynau	0.33-7.5	600-5400 @ 1800 input	0.1	85	(3:1-1:2) 6:1 total) (3:1-1:3) 9:1 total
Nutating cone	Vadetec	15-150	0-1800 @ 1800 rpm		90-94	0-1:1 (reversible unit on test)
Free ball	Contraves	0.25-2	-0.8 to +0.8 times input speed	1	85-90	0-1.25:1
		0.2-5		1	85-90	(10:1-1:2 20:1 total
	Heynau	0.125-0.5	600-5400 @ 1800 input	0.1	85	(3:1-1:2) 6:1 total (3:1-1:3) 9:1 total

Change ratio at rest	Date introduced U.S.	Date introduced Other	Additional Gearing	Mounting Foot	Mounting Flange	Operating position Up	Operating position Down	Operating position Hor.	Rotation bi-directional	Other
No	1971	1949	Yes	Yes	Yes			Y	Yes	⎫
No	1977	1947	Yes	Yes	Yes	Y	Y	Y	Yes	⎪ Either shaft may be used as the input; speed ranges shown are for the normal input and output shafts
No	1954		Yes	Yes	Yes			Y	Yes	⎬
			Yes	Yes	Yes			Y	Yes	⎪
			Yes					Y	Yes	⎪
No	1974	1959	Yes	Yes	Yes	Y	Y	Y	Yes	⎭
Yes	1975	1970	Yes	Yes	Yes	Y	Y	Y	Yes	
			Yes	Yes	Yes	Y	Y	Y	Yes	
No			Yes	Yes	Yes			Y	Yes	
No	1972	1936	Yes	Yes	Motor only	a	a	Y	Yes	
No	1972	1936	Yes	Yes	Motor only	a	a	Y	Yes	
Yes	1935		Yes	Yes	Yes	a	a	Y	Yes	
No	1972	1960	Yes	Yes	Yes	b	b	Y	Yes	New 0-1000 rpm unit @ 1750 rpm input available
No	1972	1960	Yes	Yes	Motor only	b	b	Y	Yes	
No			No	Yes				Y	No	
Yes			No	Yes	Motor only			Y	Yes	
Yes	1977	1936	Yes	Yes	Motor	Y	Y	Y	Yes	With planetary gearing can go to 0 and reverse
Yes	1971		Yes[c]	Yes	Yes			Y	Yes	
Yes	1973	1963		Yes				Y	Yes	
Yes	1973	1963		Yes				Y	Yes	
Yes	1977	1963	Yes	Yes	Motor only	Y	Y	Y	Yes	

(continued)

Traction Drive Characteristics (continued)

Type	Supplier	Horsepower rating	Speed range (rpm)	Speed change at load (%)	Efficiency (%)	Ratio
[Free ball]	Floyd	1	-1 to +1 times input speed	3-5	75	0-1:1
Disc						
Beier	Sumitomo	0.5-200	350-1400 @ 1750 input	2-4	75-87	3.3:1, 4:1
Planetary	Shimpo	3-20	0.2-1 times input speed	1-2	85	5:1
	Simplana/Lenze	0.33-25	200-1200 @ 1800 input	3-7	75-84	6:1
Toroidal	Arter	0.2-14.75	215-1700 @ 900 input 200-2000 @ 900 input	Speed ranges vary with size	To 95	(4.18:1-1:1.88) 7:1 total to (3:1-1:3.33) (4.5-1:2.22) 10:1 total (ratios depend on unit size)
		0.2-7.37	0-2200 @ 900 input, built-in planetary		To 95	0-1:2.44
	Metron	0.025	350-8400 @ 1750 input	9	70	25:1
		0.016	-1750 to +1750 @ 1750 input with planetary	5	65	0-1:1
	David Brown Sadi	0.33-5.36	540-3780 @ 1750 input	3	To 80	(3.24:1-1:2.16) 7:1 total
	Excelermatic	220	590-5250 @ 1750 input	Selectable 0.25-10	To 96	(3:1-1:3) 9:1 total
Planetary Fixed ratio	Wedgtrac	0.086-216	To 33,250 @ 1750 input; & up to 70,000 depending on input	0.5	+90	Various fixed ratios 3:1-20:1
	Nastec	10-500	Up to 480,000 depending on input		94	Various fixed ratios up to 150:1
	Paxton					
Variable ratio	Chery/Fafnir	To 150	Max. operating 5000		80 to +93	4:1
			-1750 to +1750		80 to +93	-1 to +1 (regenerative)
			0-1750		80 to +93	0-1:1[f] (regenerative)
Spool	Excelermatic (TPI)	0.2-15; 50	700-4400 @ 1750 input	1-2	85 to 90	(2.5:1-1:2.5) 6.25:1 total
Offset sphere	J. R. Young	To 20	425-7000 @ 1750 input	Very low	To 95	(4:1-1:4) 16:1 total 20:1

[a]Special.
[b]Specify.
[c]Vary output gear sets.
[d]Only to lower speed.
[e]Speed ratio controlled mechanically or automatically by load.
[f]Automatic lockup at 1:1 ratio.

Appendix

Change ratio at rest	Date introduced U.S.	Date introduced Other	Additional Gearing	Mounting Foot	Mounting Flange	Operating position Up	Operating position Down	Operating position Hor.	Rotation bi-directional	Other
Yes	1965			Yes	Motor only			Y	Yes	
No	1966	1952	Yes	Yes	Yes		Y	Y	Yes	
No		1972	No	Yes	Motor only		Y	Y	Yes	
Yes[d]	1964	1957	Yes	Yes	Yes	Y	Y	Y	Yes	
Yes		1936	Yes	Yes	Yes	b	b	Y	Yes	
Yes			Yes	Yes	Yes	b	b	Y	Yes	
Yes	1954			Yes				Y		
Yes	1965			Yes				Y		
No	1965		Yes	Yes	Yes	a	a	Y	Yes	
No			Yes					Y	Yes	
No	1973		No	Yes				Y	No	Can be used as a reducer or increaser (i.e., either shaft as input)
No	1960s		No		Yes			Y	Yes	Can be used as a reducer or increaser (i.e., either shaft as input)
e			Yes	Yes	Yes			Y	Yes	
e			Yes	Yes	Yes			Y	Yes	
e			Yes	Yes	Yes					
No	1971			Yes	Yes			Y	Yes	Max. input 100,000 rpm
No	1973		Yes	Yes	Yes			Y	Yes	Also a version to go through 0

4

BASICS AND SIMPLE FORMULAS

4.1 FUNDAMENTAL UNITS: CUSTOMARY SYSTEM

The fundamental units of any engineering system are mass, length, and time. In the customary system used in the United States these are the pound, foot, and second. Because we use the same word for mass as we do for force, we admit a certain amount of confusion into our system.

The acceleration of gravity, g, is 32.174 feet per second per second (ft sec^{-2}) at the earth's surface at sea level. Mass, which is invariant with distance from the earth's surface, is related to the force necessary to support or constrain it by Newton's famous formula

$$F = ma \qquad (4.1)$$

where

F = force
m = mass
a = acceleration

in any consistent set of units. In the customary system F is measured in pounds and a is measured in feet per second per second.

The units in which mass are measured are described as follows:

$$m = \frac{F}{a}$$
$$= \frac{lb}{ft\ sec^{-2}}$$
$$= lb\ sec^2\ ft^{-1}$$

Fundamental Units: Customary System

A mass that weighs 1 lb on a scale at the earth's surface has the value

$$m = \frac{1 \text{ lb}}{32.174 \text{ ft sec}^{-2}}$$
$$= 0.031081 \text{ lb sec}^2 \text{ ft}^{-1}$$

It follows that a mass that weighs 32.174 lb on a scale at the earth's surface would have the value

$$m = \frac{32.174 \text{ lb}}{32.174 \text{ ft sec}^{-2}}$$
$$= 1 \text{ lb sec}^2 \text{ ft}^{-1}$$

A mass this size is called a slug. The slug is not a fundamental unit, but a derived one, since it can be expressed in terms of the more fundamental pound, foot, and second. The slug is not used in this text.

Many people try to sort out the difference between pound force and pound mass by consistently using lb(f) or lb(m) in their notations. The present authors prefer never to recognize lb mass. All references to "lb" will be to force. Some equations require mass to be used to be correct. In that case the m will be written in the equation but the value

$$\frac{\text{weight}}{g}$$

will be substituted for the mass term. (The abbreviation "lb" for pound comes from the Latin "libra," which means "scale." Persons born between September 23 and October 22 will recognize Libra as the seventh sign of the zodiac.)

A units check (dimensional analysis) is a very powerful tool to determine whether one has gotten the units for mass or weight right, or for that matter whether the format of any equation is valid.

For example, consider the following equation for the natural frequency of a simple spring-mass system:

$$w_n = \left(\frac{k}{m}\right)^{1/2}$$

where

w_n = natural frequency, rad sec^{-1}
k = spring rate, lb ft^{-1}
m = mass in appropriate units (see below)

Square both sides of the expression to make it easier to examine:

$$w_n^2 = \frac{k}{m}$$

Substitute units for the variables:

$$\sec^{-2} = \frac{\text{lb ft}^{-1}}{\text{mass}}$$

One can quickly see that if mass were mistakenly entered as pounds, then

$$\sec^{-2} = \text{ft}^{-1}$$

which is an absurdity.

If mass were correctly entered as weight/g, then

$$\sec^{-2} = \frac{\text{lb ft}^{-1}}{\text{lb}/(\text{ft sec}^{-2})}$$

$$= \frac{\text{lb ft}^{-1}}{\text{lb ft}^{-1} \sec^2}$$

$$= \sec^{-2}$$

which is correct.

Dimensional analysis will allow you to judge whether you have the format of the equation correct. It will not tell you whether coefficients or other dimensionless terms are in the right positions.

4.2 DERIVED UNITS

4.2.1 Torque

Torque is a twisting effect about an axis. It is expressed as a force times the perpendicular distance from the axis to the line of action of the force. Thus torque is the product of pounds and feet and will be expressed as pounds feet. The authors will always place the force unit first when describing a torque. The force unit need not be pounds and the distance unit need not be feet. Ounce inches or pound inches are perfectly acceptable methods of describing torque as long as the torque is a product of these two units and the length dimension is described as the distance to an axis perpendicular to the force line of action. Figure 4.1 illustrates this relationship. Note that the distance measure may be called the moment arm.

Derived Units 123

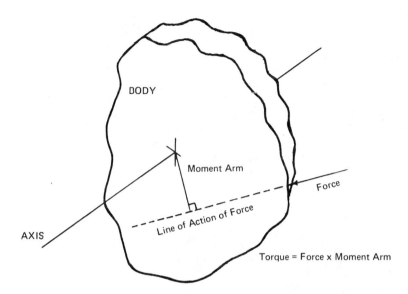

FIG. 4.1 Definition of torque.

4.2.2 Work

Work is accomplished in the engineering sense when a force is moved through a distance. The units of work are foot pounds. The authors will use the length dimension first whenever describing work. This is in contrast to the notation for torque described above. A very important difference between these similar-sounding terms is that the force must *move* in order to do work. The distance dimension is in the same direction as the line of action of the force, not at right angles to it as in the case of torque. If you picked up an object that weighed 1 lb and lifted it 3 ft, then carried it horizontally 10 ft, the amount of work done would be 3 ft lb. During the pickup the force being exerted was in the same direction as the object was moving. During the carry the force is vertical but the motion is horizontal, so no further work is being done. When torque is exerted on an object that does not turn, or when force is exerted on an object that does not move, no work is done.

When an object on which torque is being exerted does move in response to a torque, then the distance through which the force travels to do work is along the circumference of an imaginary circle of which the moment arm is the radius. That is why 1 lb ft of torque does 2π foot pounds of work on a shaft that it turns through one revolution.

4.2.3 Power

Power is the rate of doing work. One horsepower is the equivalent of 550 ft lb of work done in 1 sec, or 33,000 ft lb of work done in 1 min.

When a motor turns a shaft against a resistance, work is being done by the torque. Think of the torque as a tangential force at the end of a 1-ft arm (Fig. 4.2). When a force F turns the shaft one revolution, it has traveled 2π feet (albeit in a circle) and has done $2\pi F$ foot pounds of work. The torque T is numerically equal to the force F when the moment arm for F is 1 ft.

The horsepower of a motor is

$$hp = \frac{2\pi TN}{33,000}$$

where T is the torque in pound feet and N is the number of revolutions per minute, which is recognized as the familiar horsepower formula

$$hp = \frac{TN}{5252} \qquad (4.2)$$

Sometimes the formula is written

$$hp = \frac{TN}{63,025} \qquad (4.2a)$$

where torque is expressed in pound inches.

Power has a directional sense. When a torque is exerted on a shaft that is turning, power is being delivered to the shaft if the direction of applied torque is the same as the direction of rotation (Fig. 4.3). Power is being accepted from the shaft if the torque exerted on the shaft is in the opposite direction to the direction of rotation.

4.2.4 Energy

Energy reflects the ability to do work. It is like a bank account which records how much has been put into a system subject to later withdrawals.

A moving mass has kinetic energy, reflecting the amount of work required to get it up to its present state of motion. The datum would be the energy the same mass would have at zero velocity (zero). Presumably the mass could deliver that energy to any load that opposes its motion.

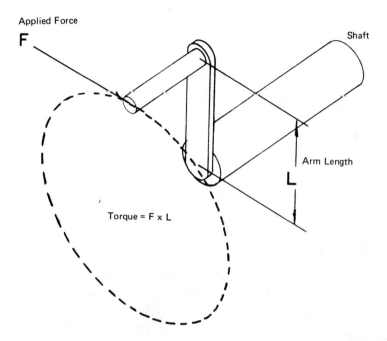

FIG. 4.2 Work being done by torque. The path length followed by the force F during one rotation of the shaft is $2\pi L$. Work = $F \times 2\pi L = 2\pi T$.

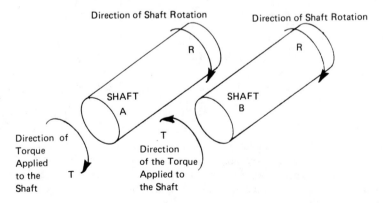

FIG. 4.3 Directional sense of power flow. Shaft A is accepting power from a source. Shaft B is delivering power to a load.

The units of kinetic energy are foot pounds. The formula for the kinetic energy of any moving mass is

$$KE = \tfrac{1}{2}mV^2 \qquad (4.3)$$

where

$$m = \frac{\text{weight}}{\text{acceleration of gravity}}$$

$$= \frac{\text{lb}}{\text{feet sec}^{-2}}$$

V = velocity, ft sec^{-1}

If the mass is spinning on an axis, such as a disc on a shaft, then it has kinetic energy by virtue of the rotation. Every point on the mass that is a distance from the axis has a velocity associated with it which is a function of the distance from the axis (radius) and the angular velocity of the mass. Each little element has kinetic energy in accordance with formula (4.3). These elemental kinetic energies can be summed up by integration, yielding

$$KE = \tfrac{1}{2} I \omega^2$$

where I is the moment of inertia (see Section 4.2.5 and Table 4.2) and ω is the angular velocity in rad sec^{-1}.

Kinetic energy, whether computed by equation (4.3) for a moving mass or by equation (4.3a) for a spinning mass, will be expressed in foot pounds or in any comparable distance × force units in other systems.

A mass that is not moving could have potential energy by virtue of its height above a datum. A rock poised above a cliff represents substantial potential energy with respect to the bottom. The units of potential energy are foot pounds. The formula for potential energy is

$$PE = wh \qquad (4.4)$$

where w is weight in pounds and h is height in feet.

Most forms of energy are convertible from one form to another. A pendulum, for example, spends its time converting the potential energy represented by the mass of the bob at the high point of its swing, to kinetic energy represented by the velocity of the bob at the low point, and back again.

All energy we know about is directly convertible into heat energy. The converse is not true. Heat energy can only be partially converted

Derived Units 127

into the other forms. This recognition has led to many philosophical treatises about where it will all end, but they are beyond the scope of this book.

The unit of heat energy is the British thermal unit (Btu). It is the amount of heat necessary to raise 1 lb of water 1°F. It is equivalent to 778.26 ft lb of mechanical energy.

Heat energy can be stored in a mass by virtue of a temperature rise. A property of materials, learned mostly empirically by experiment, is its specific heat c. Specific heat is defined as the number of Btu required to heat 1 lb of that substance 1°F.

It follows from that definition that the relationship among the factors that describe heat energy is

$$Q = cw \Delta t \qquad (4.5)$$

where

Q = heat energy, Btu
w = weight, lb
c = specific heat, Btu/(lb °F)
Δt = temperature rise, °F

Table 4.1 describes the specific heat of common engineering materials.

4.2.5 Inertia

Moment of inertia, as used here, is a term used in an angular system of units to describe the distribution of its mass about its axis of rotation. The inertia of any device can be calculated by dividing it into small blocks of matter. The weight of each block is then multiplied by the square of the distance from the center of mass of the block to the axis of rotation. The sum of all these products is the moment of inertia of the body expressed in lb ft^2. Table 4.2 describes the moment of inertia of some common geometric solids.

Frequently, moment of inertia is described as Wk^2. If the moment of inertia of any body is known, and its weight is known, then dividing the moment of inertia by the weight yields a term known as k^2. The k stands for the radius of gyration of the body. In effect, k is a distance characteristic of the mass distribution, which when squared and multiplied by the weight of the body will yield the inertia. A flywheel, which would have most of its mass located near the rim, would have a k very nearly equal to the rim radius, whereas a uniform disc of the same thickness and diameter would have a smaller value for k by virtue of its more central mass distribution compared to the flywheel.

TABLE 4.1 Specific Heats of Some Common Engineering Materials[a]

Material	Specific heat[b] (Btu lb^{-1} °F^{-1})	Density (lb ft^{-3})
Aluminum	0.21	168
Brass	0.094	526
Copper	0.10	555
Iron	0.11	486
Zinc	0.095	443
Machine oil	0.40	51.2
Water	1.00	62.4

[a]Specific heat changes with temperature. The values given are for the range 70 to 212°F.

The specific heat of principal alloys are about the same as for the base material. Therefore, most steels can be examined using the specific heat for iron.

It is a curious fact that the lighter elements have higher specific heat than the heavier ones. Water, which has a lot of hydrogen (specific heat 3.4) in it, has a specific heat nine times greater than iron.

[b]The specific heat figures, although expressed in Btu lb^{-1} °F^{-1}, are the same as the metric equivalent cal g^{-1} °C^{-1}. Water is 1.00 in both systems. The heat unit in each system was defined as that amount of heat which would raise the temperature of 1 weight unit of water 1 degree on the appropriate scale. As a result of this definition, 1 Btu lb^{-1} °F^{-1} exactly equals 1 cal g^{-1} °C^{-1}.

When a spinning mass is being driven by a motor through a speed ratio, the driver "sees" an inertia which is different from the actual inertia of the spinning mass. Consider a disc or flywheel being driven at 900 rpm by an 1800-rpm motor through a 2:1 step-down gearbox. If the system were to change speed suddenly, the amount of speed change of the flywheel would be exactly half of the speed change of the driver, because of the 2:1 ratio in the gearbox. The gearbox multiplies the torque available to change the flywheel speed by the same factor of 2. Thus the gear ratio appears twice, first by reducing the required speed change, then by increasing the torque available to accomplish it.

When a system that has shafts operating at different speeds is being analyzed, it is often convenient to "reflect" the various inertias to a common point, say the motor shaft.

Derived Units 129

$$I_{reflected} = I_{mass_1}\left(\frac{\text{speed of mass}_1}{\text{speed of motor}}\right)^2$$
$$+ I_{mass_2}\left(\frac{\text{speed of mass}_2}{\text{speed of motor}}\right)^2 + \cdots$$

Simply stated, if two masses are separated by a gear ratio, the inertia of one mass, reflected to the shaft of the other, is

$$I_{reflected} = \frac{I_{mass}}{(\text{gear ratio})^2}$$

4.2.6 Acceleration

When speed is changed in a drive system a transient torque is required during the acceleration process to store or remove the kinetic energy represented by the speed change.

TABLE 4.2 Mass Moment of Inertia of Important Engineering Solids[a]

Solid	Mass moment of inertia, I	Weight, W	Radius of gyration squared, k^2
Cylinder of radius r and thickness t	$\frac{\pi r^4 t \rho}{2}$	$\pi r^2 t \rho$	$\frac{r^2}{2}$
Right circular cone of base radius r and height h	$\frac{\pi r^4 h \rho}{10}$	$\frac{\pi r^2 h \rho}{3}$	$\frac{3r^2}{10}$
Sphere of radius r	$\frac{8\pi r^5 \rho}{15}$	$\frac{4\pi r^3 \rho}{3}$	$\frac{4r^2}{10}$

[a] ρ is the density in lb ft^{-3} when the radius r is measured in feet. The units of I will be lb ft^2. The units of k^2 will be ft^2. Other consistent units will produce results correct in those units.

Results for more complex shapes can be obtained by superposition. For example, a hollow cylinder of outside radius r_o and inside radius r_i can be calculated by subtracting the inertia of the cylinder r_i from the inertia of the cylinder r_o. Weight can be done in the same way. Then the radius of gyration squared can be calculated by dividing the net inertia by the net weight.

The governing relationship is

$$T = I\alpha \qquad (4.6)$$

where

    I = moment of inertia
    α = angular acceleration, the rate of speed change, rad sec^{-2}
    T = torque

all in consistent units. If inertia is expressed in lb ft^2, it must also be divided by the acceleration of gravity g, 32.17 ft sec^{-2}, in order to have the torque stated in lb ft.

In many kinds of equipment, the torque needed to get the equipment up to speed in a specified amount of time may size the prime mover and the drive, as the steady-state operating torque may be so low that it may not be the critical factor. In a conveyer drive, for example, the main design problem may be to get it started and stopped safely, not in keeping it moving.

Acceleration may be positive or negative. A negative acceleration simply means that speed is being reduced. In some systems speed may have to be reduced more rapidly than it may have to be increased, and negative acceleration will size the structural parts. This could be true in a punch press or in any system with limited travel that approaches a solid mechanical stop. Mechanical stops are bad news for any drive train. Special provisions for load limiting have to be incorporated to prevent self-destruction if there is even a remote chance that a mechanical stop may be encountered during any part of the load cycle.

4.2.7 Efficiency

All systems that move suffer power losses because of friction between parts which have relative motion, and because of turbulence and viscous drag of disturbed fluids.

Efficiency is the ratio of system output power to system input power. It is also expressible as the ratio of the value of system output torque or other parameter compared to the value that parameter would have had if there were no losses.

It is part of the engineering folklore that a system whose efficiency is less than 50% will be irreversible (e.g., that it cannot be driven by its output shaft). This is not necessarily true. If the system has only one significant loss-producing element, such as a worm and wheel, the formula

$$\text{Efficiency}_{\text{output driving}} = 2 - \frac{1}{\text{efficiency}_{\text{input driving}}} \qquad (4.7)$$

Derived Units 131

will be valid. When the efficiency input driving term is 0.50 or less, the efficiency output driving term will be zero and the system cannot be driven by its output shaft.

If the system has many stages each of which has only a small loss, each stage can be driven by its own output shaft. The system, which consists of a train of such stages, can be back driven from the output even though the overall efficiency, taken as the product of the stage efficiencies, is below 0.50.

One of the sad things in engineering is that you cannot count on friction to hold a system locked in this manner. An "irreversible" system will happily reverse in the presence of severe vibration.

4.2.8 Recirculating Power

The concept of recirculating power is especially important in differential-type drives, which can go to zero output speed while carrying a substantial load. The point of particular interest is that the power *level* in some of the shafts of the drive may be much greater than either the input power or the output power. This does not violate any rules, but it can do startling things to efficiency.

One of the clearest ways of showing the concept of recirculating power is by examination of a degenerate form of the differential chain hoist. Consider Fig. 4.4; this chain hoist has two pulleys which are shown equal in diameter. Normally, differential hoist pulleys are bolted together. For purposes of this discussion they are shown separated, but each is mounted securely to a common shaft. A bridge picks up the common shaft and hooks it to a beam in the ceiling.

A heavy load is supported on a pulley which is suspended from a loop in the chain. The chain is threaded over one shaft pulley, through the load pulley, forming a loop, and back over the other shaft pulley. The direction of chain passage over both shaft pulleys is the same. Therefore, the chain passes diagonally underneath the shaft. The tension in the chain parts due to the load tends to rotate the front pulley clockwise and to rotate the rear pulley counterclockwise. The torque is balanced through the connecting shaft. The chain forms a slack loop back to the starting point. The pulleys, of course, have dimples or cogs which engage the chain and prevent it from slipping.

Assume that the pulleys each have a radius of 1 ft. If a load of 1000 lb were placed on the pulley suspended in the loop, there would be 500 lb of tension in each of the chain parts which carry the load back to the front and rear pulleys. The front and rear pulleys each will have a torque of

$$500 \text{ lb} \times 1 \text{ ft} = 500 \text{ lb ft}$$

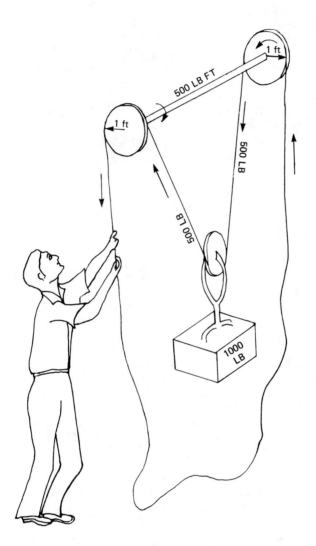

FIG. 4.4 Degenerate form of differential chain hoist. (Drawn by Gail Tarkan.)

Derived Units 133

There will be no tendency for either of the pulleys to turn because the torque loads placed on the connecting shaft by the pulleys are equal and opposite. (In a normal differential hoist the pulleys would be of slightly different diameter. This degenerate form, using equal-diameter pulleys, was invented just for this example.)

If a man pulls on the slack chain at, say, 5 ft sec^{-1}, he does no useful work against gravity, for the load does not rise. The chain simply reeves through the pulleys. There is no shortening of the loop because for each foot of chain taken up by the front pulley, exactly 1 ft of chain is paid out by the rear pulley. If the system has no friction or other losses, the man pulling the chain does no work. His output horsepower is zero.

Consider the shaft. It experiences a torque of 500 lb ft from each end. Since the man is pulling the chain at 5 ft sec^{-1}, the pulleys rotate at

$$N = \frac{5 \text{ ft sec}^{-1} \times 60 \text{ sec min}^{-1}}{2\pi}$$

$$= 47.75 \text{ rpm}$$

Each pulley experiences a horsepower level of

$$\text{hp} = \frac{TN}{5252} \tag{4.2}$$

$$= \frac{500 \times 47.75}{5252}$$

$$= 4.55$$

If the man is pulling down on the chain in Fig. 4.4, both pulleys and the shaft are rotating counterclockwise. The torque exerted by the front pulley on the shaft is clockwise and the torque exerted by the rear pulley on the shaft is counterclockwise. Therefore, according to the notation of Section 4.2.3, the rear pulley is supplying power to the shaft and the front pulley is absorbing power from the shaft.

The horsepower in the part of the chain under tension is

$$\text{hp} = \frac{500 \text{ lb} \times 5 \text{ ft sec}^{-1}}{550 \text{ ft lb sec}^{-1} \text{ hp}^{-1}}$$

$$= 4.55 \text{ hp}$$

This is pure recirculating power! The 4.55 hp in the chain is transferred to the rear pulley. This allows the chain to go slack as it enters the return loop. The rear pulley transfers the power to the front pulley, which transfers the power to the chain, restoring the tension.

This absurd machine provides an example of recirculating power. The machine cannot run on its own because any real implementation of it will have losses. The chain has friction losses around the pulleys as the links bear on the cogged walls. The shaft has losses in its bearings. The load pulley has losses in its bearing.

These losses are very real and are proportional to the amount of mechanical power being handled. The shaft and bearings cannot tell that they are in a useless loop. They feel the same torques and experience the same losses as if they were transferring real power from a source to a load.

Consider a traction drive which had, say, 90% efficiency in its power path from motor to load, so that if it were accepting 100 hp from a motor it would have 10-hp losses and would deliver 90 hp to the load. Suppose that the load path within this drive were folded back by means of a differential so that a portion of the output torque were fed back to the input (Fig. 4.5). Power flows from the input (left) to the output (right) along the path marked by the heavy line. The direction of rotation of the output is opposite to that of the input because of the spur gear mesh at the end of the input shaft. Power flows back from the output shaft to the input along the path marked by the thin line.

Using the notation

N_1, number of teeth in the spur gear at the end of the input shaft
N_2, number of teeth in the gear which mates with N_1
N_3, number of teeth in the planetary ring internal gear
N_4, number of teeth in each planet gear
N_5, number of teeth in the sun gear
R_V, variator step-up ratio

the output ratio R_O will be

$$R_O = \frac{N_2}{N_1} \times \frac{N_3 + N_5}{N_3} - R_V \times \frac{N_3 + N_5}{N_5}$$

and the output speed will be zero when

$$R_V = \frac{N_2}{N_1} \times \frac{N_5}{N_3}$$

Such a drive could be designed so that at some position of its control it would go to zero output speed while maintaining full output torque. Its output power would go to zero but its losses would not,

Derived Units

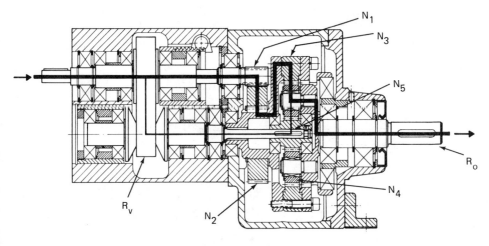

FIG. 4.5 Schematic of traction drive with folded-back power path. (Courtesy of Plessey Dynamics Corporation, Hillside, N.J.)

for most of the machinery would be running at full speed in one direction or another at full output torque.

Recirculation of power enables some drive systems to obtain very high reduction ratios with reasonably sized components. Whenever a candidate drive uses recirculating power in this way, its efficiency and heat rejection should be confirmed as satisfactory for the intended purpose.

4.2.9 Viscosity

Viscosity is a concept based on the force required to shear a fluid at a certain rate. Consider a cylinder 1 ft in circumference nested closely within but not touching another cylinder (Fig. 4.6). Consider the thin space between them as being filled with a fluid, say water, which wets and clings to the walls.

When the inner cylinder is rotated, the water film is sheared. A tangential force, tending to retard the moving cylinder, is developed due to viscous drag. The force F divided by the area A of the rotating cylinder, divided by the speed V of the moving surface, multiplied by the spacing S between the cylinders, multiplied by the acceleration of gravity G, is the dynamic viscosity of the fluid.

$$\text{Dynamic viscosity} = \frac{F \times S \times G}{A \times V}$$

$$= \frac{\text{lb} \times \text{ft} \times \text{ft sec}^{-2}}{\text{ft}^2 \times \text{ft sec}^{-1}}$$

$$= \text{lb ft}^{-1} \text{ sec}^{-1}$$

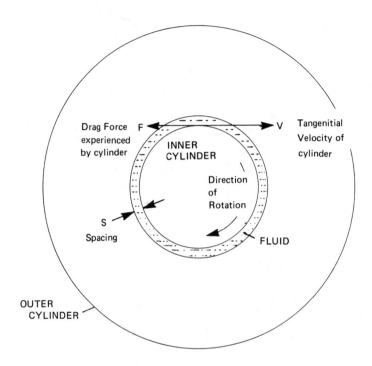

FIG. 4.6 Viscosity measurement.

The units of dynamic viscosity in the customary system are usually lb ft^{-1} hr^{-1}. Viscosity determined by the test apparatus described above would be multiplied by 3600 sec hr^{-1} so that the results could be published in terms of lb ft^{-1} hr^{-1}.

Dynamic viscosity may be converted to centipoise by multiplying the lb ft^{-1} hr^{-1} value by the appropriate conversion factors from Table 4.3. One lb ft^{-1} hr^{-1} is equal to 0.00413379 g sec^{-1} cm^{-1}. One g sec^{-1} cm^{-1} is a poise, named after J. L. M. Poiseuille, who investigated viscosity in 1842 by observing and calculating the flow rate of various viscous fluids through capillary tubes. The viscosity unit most often reported is the centipoise (cP), which is one hundredth of a poise. The viscosity of water at 20°C (68°F) is almost exactly 1 cP.

The preferred unit for dynamic viscosity in the Système International (SI) is the pascal second. One centipoise is 1000 pascal seconds. The pascal second is almost never used.

Viscosity is commonly determined by experiment. When the experiment is done by allowing liquids to flow by gravity through a vertical tube, then if the height of the tube is the same for all experi-

TABLE 4.3 Viscosity of Some Common Fluids

Fluid	Temperature (°C)	Viscosity[a] (cP)
Acetic acid	20	1,219
Acetone	20	0.3225
Benzene	20	0.649
Castor oil	20	621
Ethyl alcohol	20	1.192
Ethylene glycol	20	1.0
50/50 water mix	20	3.0
Formic acid	20	1.782
Fuel oil	25	515.
Glycerin	25	735.
	20	830.
Hydrogen peroxide	20	1.293
Kerosene	20	2.375
Light machine oil	20	92.
Mercury	20	1.6
Methyl alcohol	20	0.591
Olive oil	20	80.8
Sulfuric acid	20	24.2
Toluene	20	0.586
Turpentine	20	1.46
Water	0	1.792
	20	1.005
	100	0.284

[a]
Multiply:	by:	to obtain:
lb hr^{-1} ft^{-1}	0.000277778	lb sec^{-1} ft^{-1}
lb hr^{-1} ft^{-1}	0.00000863359	lb sec ft^{-2}
lb hr^{-1} ft^{-1}	1.48816	kg hr^{-1} m^{-1}
lb hr^{-1} ft^{-1}	0.00413379	g sec^{-1} cm^{-1}
lb hr^{-1} ft^{-1}	0.413379	cP
lb sec ft^{-2}	47.8803	N sec m^{-2}
lb sec ft^{-2}	47,880.3	cP
cSt	1 × 10^{-6}	m^2 sec^{-1}
ft^2 sec^{-1}	92,903.	cSt
ft^2 sec^{-1}	0.092903	m^2 sec^{-1}

ments, the pressure that drives the flow of the liquid being evaluated is proportional to the density of the liquid. Viscosity normalized by dividing lb ft^{-1} hr^{-1} by the density of the fluid is called kinematic viscosity.

$$\text{Kinematic viscosity} = \frac{\text{dynamic viscosity}}{\text{density}}$$

$$= \frac{\text{lb ft}^{-1}\text{ hr}^{-1}}{\text{lb ft}^{-3}}$$

$$= \text{ft}^2 \text{ hr}^{-1}$$

Ft^2 hr^{-1}, when multiplied by the appropriate conversion factors, can be expressed as m^2 sec^{-1}. The value 10^8 m^2 sec^{-1} is called a stoke, named after G. C. Stokes, who investigated viscosity in terms of particles settling in a fluid. The common engineering unit is the centistoke (cSt), which is equivalent to 10^6 m sec^{-1}.

Table 4.3 describes the viscosity of several technical fluids and gives conversion factors for the units. Note that in most fluids viscosity is very sensitive to temperature. Viscosity tabulations must state the temperature for which the value is given.

Fluids also exhibit changes of viscosity with pressure. Most tabulations refer to atmospheric pressure measurements only. Fluids in which the viscosity rises sharply at very high pressure are desirable for traction use (see Section 5.1).

4.3 METRIC SYSTEM

The ISO [1] has put forward a set of standard metric units, called SI (Système International) that, hopefully, all countries, whether presently metric or not, will adopt. Judging from the literature and specifications which continue to arrive from countries using the metric system, the hoped-for unity has not yet been achieved. The engineer must be prepared to convert several different metric units for force, torque, and pressure to interpret them in terms of the common units in current use in the United States.

The fundamental units for mass, length, and time are the kilogram, meter, and second, respectively. The kilogram is a standard unit of mass that would weigh 2.2046 lb at rest at the equator at sea level. The meter is a standard unit of length equal to 39.37 in. All other units are derived in some way from these three fundamental units.

Metric System 139

4.3.1 Force

Metric system users do not use the word "kilogram" for force. SI prefers the newton (N). The newton is that force which will accelerate a kilogram mass (kg) 1 meter per second per second. It is equal to 0.224809 lb. The acceleration of gravity is 9.80665 m sec^{-2}. When the pound equivalent of the newton is multiplied by 9.80665, the pound equivalent of the kilogram will be found:

$$9.80665 \times 0.224809 = 2.2046$$

Not all metric countries use the newton as the force unit. Some use the kilopond, abbreviated kp. The kilopond is the force exerted by gravity on a kilogram mass at rest at the equator. It is equal to 2.2046 lb. One will occasionally encounter the term "kilogram force, kg(f)" for this same unit. Some countries use the decanewton, abbreviated daN, which is 10 times the size of the newton, or 2.24809 lb. Table 4.4 summarizes these force conversions.

4.3.2 Torque

Torque is a force times a moment arm. The moment arm is measured at right angles to the line of action of the force, as described in Section 4.2.1. As also described in Section 4.2.1, the force dimension will always be written first when describing torque. Metric torque units, and factors for converting to their equivalents in U.S. conventional units, are listed in Table 4.5.

4.3.3 Pressure

Pressure is a force divided by an area. The pascal is the pressure of 1 newton per square meter. That unit is so small that the kilopascal (1000 N m^{-2}) is the preferred unit. Table 4.6 lists several metric pressure units which are likely to be encountered, and factors for converting to their conventional equivalents.

TABLE 4.4 Metric Conversions: Force Units

To convert:	(abbreviated)	to:	multiply by:
newtons	N	pounds	0.22481
decanewtons	daN	pounds	2.24809
kilograms force	kgf	pounds	2.20462
kiloponds	kp	pounds	2.20462

TABLE 4.5 Metric Conversions: Torque Units

To convert:	(abbreviated)	to:	multiply by:
newton meters	Nm	pound feet	0.73756
decanewton meters	daNm	pound feet	7.3756
kilogram centimeters	kgcm	pound feet	0.07233
kilogram meters	kgm	pound feet	7.2330
kilopond centimeters	kpcm	pound feet	0.07233
kilopond meters	kpm	pound feet	7.2330

4.3.4 Power

The metric unit for power is the watt, which is used for both mechanical and electrical power. A watt is equal to 1 joule per second. This is mechanically equivalent to 1 N moving through 1 m per second, or, in conventional units, to 0.73576 ft lb sec^{-1}. Because the watt is such a small unit, the kilowatt is preferred. Table 4.7 lists several metric power units likely to be encountered. The cheval vapeur, or metric horsepower, is similar to the conventional horsepower but not identical to it. It is defined as 4500 meter kiloponds per minute.

4.3.5 Temperature

The metric unit for temperature is the degree Kelvin (absolute temperature) or the degree Celsius. For fear of confusion with the unit "grad" (a 1/400 part of a circle), the familiar name "centigrade" has been dropped.

Both the degree Kelvin and the degree Celsius are the same-size unit as the degree centigrade. They both represent 1/100 of the

TABLE 4.6 Metric Conversions: Pressure Units

To convert:	(abbreviated)	to:	multiply by:
kilograms/mm^2	kg/mm^2	pounds per square inch	1422.26
kilopascals	kPa	pounds per square inch	0.14504
megapascals	mPa	pounds per square inch	145.04
millibars	mb	pounds per square inch	0.014504
torr		pounds per square inch	0.0192823

TABLE 4.7 Metric Conversions: Power Units

To convert:	(abbreviated)	to:	multiply by:
cheval vapeur (metric horsepower)	cv	horsepower	0.98632
kilowatts	kW	horsepower	1.34102

TABLE 4.8 Temperature Conversions

To convert:	to:	multiply by:	then add:
degrees Kelvin	degrees Rankine	1.8	—
degrees Kelvin	degrees Celsius	1.0	-273
degrees Celsius	degrees Fahrenheit	1.8	32
degrees Celsius	degrees Kelvin	1.0	273
degrees Fahrenheit	degrees Celsius	First substract 32 from degrees F, then divide the remainder by 1.8	

Comparable temperature levels for several important physical points:[a]

Physical point	°K	°R	°C	°F
Absolute zero	0.	0.	-273.15	-459.6
Mercury melts	234.	422.	-39.	-38.2
Ice melts	273.	492.	0.	32.
Water boils	373.	672.	100.	212.

[a]Kelvin and Rankine scales read the same at absolute zero. Celsius and Fahrenheit scales read the same at -40.

TABLE 4.9 Summary of Metric Conversions

To convert:	(abbreviated)	to:	multiply by:
calories	cal	Btu	0.003968
centimeters	cm	inches	0.3937
cheval vapeur	cv	horsepower	0.98632
decanewtons	daN	pounds	2.24809
decanewton meters	daNm	pound feet	7.3756
kilograms	kg	pounds	2.20462
kilograms/mm^2	kg/mm^2	pound/in.^{2}	1422.26
kilogram centimeters	kgcm	pound feet	0.07233
kilopascals	kPa	pounds/in.^{2}	0.14504
kiloponds	kp	pounds	2.20462
kilopond centimeters	kpcm	pound feet	0.07233
kilopond meters	kpm	pound feet	7.2330
kilowatt	kw	horsepower	1.34102
megapascals	mPa	pounds/in.^{2}	145.04
meters	m	inches	39.37
millibars	mb	pounds/in.^{2}	0.014504
millimeters	mm	inches	0.03937
newtons	N	pounds	0.22481
newton meters	Nm	pound feet	0.73756
torr	T	pounds/in.^{2}	0.019282

temperature difference between the freezing point of water and the boiling point of water at standard sea level pressure.

The corresponding temperature units in the conventional system which correspond to the degree Kelvin and the degree Celsius are the degree Rankine (absolute temperature) and the degree Fahrenheit, respectively. Both represent 1/180 of the temperature difference between the freezing point of water and the boiling point of water at sea level pressure. Therefore, the degree Kelvin or the degree Celsius is a larger unit than the degree Rankine or the degree Fahrenheit.

Temperature *rise* given in degrees Kelvin or degrees Celsius may be converted to temperature *rise* in degrees Rankine or degrees Fahrenheit simply by multiplying by 1.8. This is because temperature rise is a *difference* function which does not require a reference point or an actual thermometer reading.

The Kelvin and Rankine scales are used for thermodynamic calculations because they begin at absolute zero, which is located at -273.15°C or at -459.6°F. The Celsius scale has its zero placed at the melting point of ice. Gabriel Fahrenheit set the zero of his scale at the lowest temperature he could achieve with a mixture of ice and salt. He used human body temperature as the 96-degree mark. When the scale was divided into individual degrees. the melting point of ice came out at the 32-degree mark, and the boiling point of water came out at 212 degrees. We can measure body temperature more accurately today than Fahrenheit could in 1714, but his scale has remained in use as he proposed it.

It is important to reiterate that temperature has a *level* which is measured by a thermometer calibrated in degrees from an arbitrary reference and labeled according to the name of the scale which starts at that reference. The *size* of a degree Celsius or a degree Fahrenheit has nothing to do with the reference. A 40°C rise is the same as a 72°F temperature rise regardless of the starting point of the rise. The conversion of a temperature rise in one system to the equivalent temperature rise in the other is a simple matter of multiplying (or dividing, as appropriate) by 1.8.

The conversion of temperature level in one system to temperature level in another is slightly more complex because the systems start at different reference points. Table 4.8 gives the methods of converting temperature level between the different systems, and gives some important physical constants in terms of all four scales.

4.3.6 Summary of Metric Conversions

Table 4.9 summarizes all the metric conversions which may be of interest in traction drive applications.

REFERENCE

1. 150 1000-1973 E, International Standardization Organization, Geneva, Switzerland.

5
TRACTION

5.1 TRACTION FUNDAMENTALS

Traction drives, as defined in this book, are drives using hardened and accurately finished steel traction elements rolling against each other in the presence of a fluid. The rolling contacts do not squeeze all of this fluid out of the contact area. A very thin film of fluid remains which separates the metal surfaces. The film may only be 1/10,000 as thick as the contact is wide, but the fluid properties in this film determine the traction produced by the contact. This trapped fluid is very viscous due to very high compressive forces, but it is still not a solid. Surface irregularities may be large relative to the film thickness, but the compression of the metal in the contacting surfaces is often hundreds of times greater than the film thickness.

Surface irregularities have little effect in the transfer of traction force from one rolling element to another. This transfer is due largely to shear in the fluid film, and is directly proportional to the rate the film is sheared and to the film properties. It follows that since the transfer of traction forces is due to shear motion, there will be no transfer if there is no shear motion. All traction drives have small shear motions in the direction of rolling that cause a creep of the contact. The amount of this creep is a function of a great many variables, and may range from less than $\frac{1}{2}$% to 5% or more. There are several kinds of shear motions which may occur simultaneously in a traction contact. Creep is the useful shear motion which is necessary

---

This chapter is compiled from *Rolling Traction Analysis and Design* by Charles E. Kraus, published in 1972 by Excelermatic, Inc., Austin, Texas, with additional material added by the author.

to transfer the traction force in the driving direction. Because the magnitude of creep may vary with operating conditions such as speed, temperature, contact force, and load, traction drives cannot be used for timing.

Skew motions within the film occur as a function of drive geometry. These motions generate forces similar to the useful traction force, but which are not aligned with the direction of motion. Therefore, they put loads on the drive mechanism which have to be accounted for, and they result in losses that heat the fluid and the metal in the contact area. One kind of skew force would result from misalignment of traction elements as a result of design or of manufacture or of deflection. Another, more common, skew motion, called spin, occurs when all elements of the contact zone do not have identical relative motion. Spin, to some extent, is present in all variable-ratio drives since such drives run dissimilar geometric elements against each other over finite contact widths. For example, when a cone is run against a ring, all elements of the ring move equal distances in equal time, but the corresponding elements of the cone do not, since each is at a different radius.

It cannot be presumed that the shear motion due to creep will be the dominant motion in a rolling traction contact. Rotational and skew motions superimposed on a contact because of the geometry of a drive will add motions which vary in magnitude throughout a contact area. Different little elements of the area may have different shear motions which may lie in different directions, even in a direction opposite to the rolling direction. The forces generated in each little element of area are therefore different in magnitude and direction. Useful traction is the summation of these forces in the rolling direction. Losses are the summation of the losses of all the area elements.

There are two chief elements of traction drive design that affect overall drive performance. The first is geometry-produced shear. Traction forces are generated in a traction contact as a function of load on the contact. The ratio of force to load is affected by fluid properties. All the losses in a contact are converted to heat. This heat is conducted away by the metal and some of the fluid, but temperatures in the contact can reach destructive levels for the fluid if shear becomes excessive. If the drive geometry is such that the rotational and skew shear components are large, the only way to hold losses within a safe range is to limit the load and the traction forces. Actually, all drives must do this to some extent, but designs that control geometry-produced shear can be rated for larger contact forces.

The second important design element is provision for adequate contact force. Insufficient contact force for the needed traction force will increase the required creep rate and may cause uncontrolled contact creep, or gross slip. Some drives are designed deliberately with light contact loading to allow such slippage as a safety factor, with very little wear. However, most drives can be damaged by excessive creep and need adequate contact force. Commonly, some type of loading means is used, so contact force is a function of input or output torque on the drive. Ideally, the contact force should increase directly with the required traction force and be imposed on the contact before the traction force increases. Where this is done, heavy-duty traction drives are capable of greater shock and overloading than gearing of equivalent rating.

Normal traction drive failure is by wear or fatigue of rolling surfaces. As such, traction drive failures are benign, with ample warning, rather than catastrophic. This can be an advantage over competitive drive concepts in many applications.

5.2 TRACTION COEFFICIENT

Consider a hardened-steel sphere rolling on a flat steel plate, without lubricant, as in Fig. 5.1. The two elements are pressed together by a force normal to the plane of contact, labeled "contact force." The diameter of the sphere is labeled "diameter of rolling." Diameter of rolling is an important measurement for comparative purposes, but its effect is greatly modified by other geometric considerations. If there were no contact force on the sphere, the point of contact with the plate would be literally that, a point with no area. With a force present, an area develops because the force deflects or flattens both the plate and the sphere. In outline, this area is a circle because the chosen illustration is of a sphere in contact with a flat plate. In most traction contacts the pattern would be elliptical. The pattern diameter in the direction of rolling is the rolling direction contact diameter, and the diameter at right angles to this is the transverse direction contact

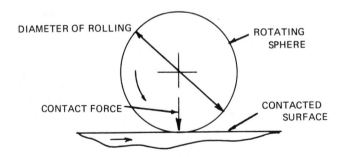

FIG. 5.1 Sphere rolling on flat plate.

Traction Coefficient

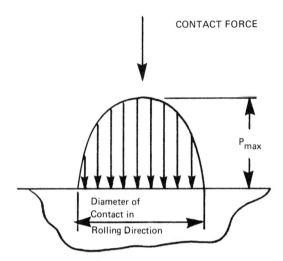

FIG. 5.2 Classical Hertz stress distribution.

diameter. The case of a sphere rolling on a plate is a special degeneration of the contact ellipse in which the transverse and rolling diameters are equal. In most cases these diameters are very different. In the special case of a cylinder rolling on a cylinder the degenerate form of the contact ellipse is a rectangle.

Most engineers are familiar with the classical Hertz stresses and contact force pattern. An in-depth discussion of these stresses is given in Chapter 7, since they form the basis for fatigue-life prediction. The pattern for the sphere on the plate is shown schematically in Fig. 5.2. Figure 5.2 is a plot of the contact force along the centerline of the contact in the rolling direction. The peak value of the diagram is 1.5 times the average value. This force pattern is found only when the contacts are not lubricated, or are stationary. In rolling traction contacts in the presence of a fluid, the pattern is somewhat different.

Figure 5.3 shows an enlarged schematic view of the real contact zone of Fig. 5.1. As Fig. 5.3 illustrates, when a fluid and rolling are both present, the two traction element surfaces are separated by a pad of fluid. True, this pad is not thick, but separation of the contact surfaces is complete if surface irregularities do not puncture the pad. The pressure profile through such a lubricated rolling contact is surprising. Pressure begins to build up before the area of classical contact is reached due to compression of the fluid. The fluid being rolled on attempts to escape on all sides and, in doing so, creates a pressure ridge in a horseshoe shape around three edges.

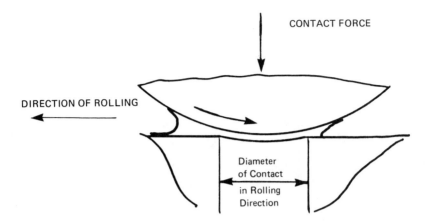

FIG. 5.3 Diagram of real contact zone with lubricant.

The pressure suddenly drops off at the end of the contact area, and may even go negative for a small spot on the centerline of rolling.

Figure 5.4 shows the effect of the factors discussed above on the classical Hertz force distribution of Fig. 5.2. As Fig. 5.4 shows, the pressure profile, aside from the effects noted above, is about the same as the classical Hertz force distribution when the contact force is high. As the contact forces decrease, the force pattern becomes flatter and the ratio of the peak force to the average force decreases. Much analytical work has been done to determine the effect of this force distribution.

The next several pages relate traction force to contact force and arrive at a coefficient, μ, which relates them under the circumstances of the discussion. The μ value obtained will be based on the average pressure across the contact area. The ratio of maximum pressure (or force) to average pressure (or force) is 1.5:1 for classical force distribution, but it varies to almost 1:1 in low mean pressure traction if pressure spikes are disregarded. In analyzing high-contact-pressure cases the classical force distribution should be used, but the error of using a uniform distribution is rarely over 10%.

Figure 5.5 shows vectorially the addition of a tangential force to the contact force between the sphere and the plate. The tangential force is the traction force which is developed at the contact. The ratio of the tangential force to the contact force is μ, the traction coefficient; μ is also the tangent of the angle that is created when the two forces are added vectorially. Operative values of μ may range from 0.01 to 0.12. Section 5.3 describes μ as a function of traction fluid, and Section 5.4 describes the effect of spin (geometry imposed shear) on μ.

Traction Coefficient   149

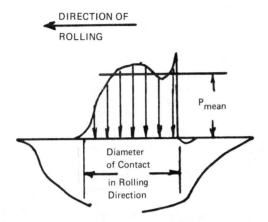

FIG. 5.4 Real contact stress pattern with lubricant.

Chapter 6 summarizes all these effects in curves that combine the variables.

Figure 5.3 showed that the traction elements were separated by a space filled with, by definition, a lubricating fluid. How, then, can a tractive pull be transmitted from one element to the other? The answer lies entirely within the fluid. The steel elements contribute and determine the area involved and, as we shall see later, they affect the fluid pad in a few important ways; but the fluid is the important traction transmitting means. It is obvious that traction, to be transmitted by a film pad, must be done by shear resistance in the fluid. Within the contact the fluid is subject to a high compressive force.

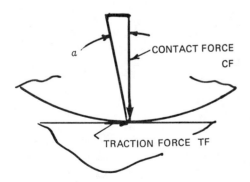

FIG. 5.5 Contact zone forces ($\mu = TF/CF = \tan \alpha$; α = traction angle).

Its viscosity is increased so greatly that some of the better fluids become virtually a plastic at these pressures and behave like a glass or a very stiff tar.

Shear rate is essentially proportional to tractive pull until heat development and excess slip put a limit on the tangential forces developed. Section 5.3 discusses these fluids in some detail, as new developments in special synthetic fluids have vastly increased the potential of traction.

We may now look at the basic type of traction curve as shown in Fig. 5.6. The traction coefficient, μ, is plotted as a function of shear rate. Figure 5.6 is drawn for purposes of this discussion, without specific values. This will also be true of Figs. 5.7-5.11. Comparable figures are developed in Section 5.3 with specific information. Chapter 6 contains an extensive treatment of specific information which is based on terms and concepts developed in Sections 5.2-5.5.

Referring to Fig. 5.6, note that for increasing shear rate the traction coefficient at first increases linearly, then rounds over into a downward-sloping line. The curve ends somewhere, depending on the fluid, because of a complete breakdown of the pad at some higher shear rate value that allows metal-to-metal contact to occur. This phenomenon is discussed in Section 5.5. Below certain pressure thresholds, depending on many factors (see Section 5.5), such contact may not cause disastrous wear. Many low power drives are designed to operate with such contact frequently present. Above this threshold pressure, such contacts tear up the surfaces, weld the metal surfaces, and destroy the temper. Obviously, a traction drive must never be allowed, by any means, to operate in that range of the curve.

The basic traction curve is sometimes carried back to zero shear rate as a dashed line, as shown in Fig. 5.6. Whereas this may be a

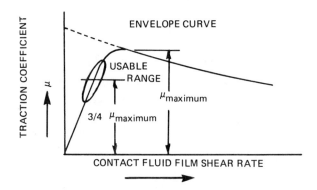

FIG. 5.6 Basic traction curve: traction coefficient as a function of shear rate.

Traction Coefficient

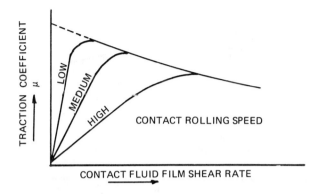

FIG. 5.7 Effect of contact rolling speed on traction coefficient.

drafting aid, it has no physical meaning. We know that the pressure pattern in the contact changes from the static distribution while rolling, and we are concerned only with rolling contacts. If shear rate goes to zero while rolling, it can only do so if there is no tangential force present. It also follows that if tangential force is present to cause shear in the film pad, then there is a value of shear rate.

In a well-designed traction drive contact force is usually a function of applied torque on the drive. The required traction force, of course, is also a direct function of applied torque. If the design is such that μ is a reasonably constant value (e.g., near the knee of the curve of Fig. 5.6), shear rate may also be reasonably constant for otherwise unaltered conditions.

The selected value for μ for design purposes must be safely below the maximum μ value. If it is too low, the drive will be subjected to needless forces, will be too large, and will lose efficiency. As shown in Fig. 5.6, there is a usable range, and it is recommended that a design value of μ be normally selected at about three-fourths of the value of maximum μ for the lowest value of mean contact pressure to be encountered (see Fig. 5.6).

One variable with a major effect on μ is rolling speed. Figure 5.7 illustrates this, showing how three curves for traction coefficient μ, as a function of shear rate, drawn for low, medium, and high rolling speed, tend to blend into the same envelope curve. μ must be determined for the highest rolling rate intended for the design. For all speeds lower than this the design will then be conservative (e.g., a higher value of μ will be available than will be required).

Figure 5.7 shows that for a given μ, shear rate tends to increase with rolling speed. Experience teaches that the two tend to increase in step such that creep, when expressed as a percentage, tends to

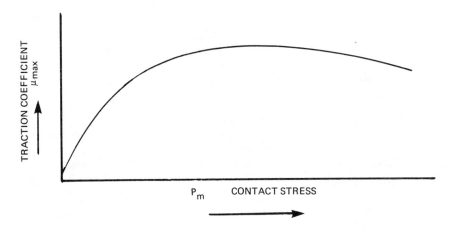

FIG. 5.8 Maximum traction coefficient as a function of mean contact stress.

be reasonably constant. The percent of creep may increase or decrease under some conditions, but it is not so far from constant that it cannot be considered constant as a rough rule of thumb.

A variable with an important effect on traction coefficient is the mean contact stress level. Figure 5.8 shows the type of curve normally encountered when μ is plotted as a function of mean contact stress P_m. The curve, of course, starts at μ equals zero for no P_m. This lower corner is of no interest except for the fact that it is the region of hydrodynamic lubrication. A traction contact is not thought of as an ideal bearing surface, but there are conditions where hydroplaning can drop a traction contact suddenly into a low μ value. One of these conditions is extreme cold, because of its effect on fluid viscosity. A drive designed for light-viscosity fluid might have trouble if that fluid turned to a grease-like consistency. The shape of this curve and its values tend to determine minimum imposed forces for safe operation. Such forces can be placed on contacts by preload means, and also by the shape and design of automatic loading devices such as loading cams.

Heavy-duty traction drives, to be used under a wide range of conditions, normally will have a relatively heavy preload and will present a noticeable drag when turned by hand. This has no effect on running efficiency or life.

Maximum values of μ tend to occur at about the same range of P_m for the most commonly used traction fluids. Although the knee of the curve is poorly defined, this P_m range is usually between 200,000 and 275,000 psi (1379 and 1896 MPa). The one exception could be the

Traction Coefficient

silicone fluids, but these have limited use, being largely supplanted by superior synthetic napthenic types. The μ-P_m relationship is considerably affected by any factor that affects the thickness of the fluid contact pad. Fluid viscosity, rolling rate, and transverse contact diameter have pronounced effects. Extreme cold was mentioned, but high temperature also has an effect. This is illustrated in Fig. 5.9.

The traction coefficient, μ, is plotted as a function of temperature range of, say, 40 to 250°F (4 to 121°C) when using a typical traction fluid. Curve A is the type usually encountered for very smooth contact surfaces. Because of surface wear problems, many small drives tend to use these very smooth surfaces. Lapping and polishing to even 4 rms has been done. With smooth surfaces μ definitely decreases at higher temperatures. When light surface wear is present, there is not much that can be done about this reduction of μ, as the wear will polish the surfaces. If a surface is not subject to wear, which would remove an applied finish, it can be given a surface favorable to traction and it is found that the droop of μ with temperature can be eliminated. Uncontrolled roughening of the traction surface will not work. Although a curve such as C can actually be produced, it has been found that there exists a threshold of surface treatment where asperities disrupt normal shear patterns and drastically reduce life.

Spin is one kind of skew shear, defined as a rotation superimposed on a rolling contact as a function of the geometry of the drive design. Spin is discussed in Section 5.4. At this point we will only indicate a little about how it tends to reduce the value of the traction coefficient.

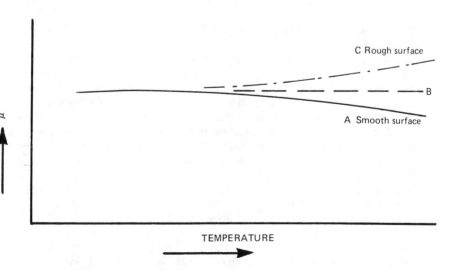

FIG. 5.9 Effect of temperature on traction coefficient.

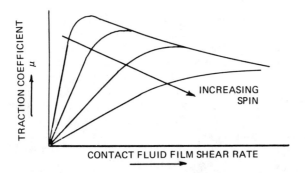

FIG. 5.10 Effect of spin on traction coefficient.

Figure 5.10 shows a family of μ versus shear rate curves for a number of spin values. The loss of usable μ with increasing spin is readily apparent.

Figure 5.10 is of the same form as Fig. 5.7, in which a family of curves of μ as a function of shear rate was plotted for increasing values of rolling speed. A full understanding of the effect of spin requires analysis of the contact area phenomena. Spin not only affects μ directly but largely controls losses and failure types. Since spin cannot take place without shear in the fluid pad, it follows that this shear adds vectorially to the shear due to the tangential force and therefore will affect the basic working curve for the design involved.

5.3 TRACTION FLUIDS

The fluid used in a traction drive has a number of different requirements. Primarily, in the rolling contact, it must be capable of transferring considerable force when being sheared in a very thin film. It is important that it resist deterioration and maintain its viscosity during the shearing. The fluid should also be a good lubricant, able to lubricate bearing cages and sliding surfaces that may be present as well as gears which are frequently built into traction drives. The lubricant is the major means of carrying the heat generated by power losses to the housing walls or other cooling surfaces. Several characteristics of the fluid may be very important, depending on the drive design and use. If there are hydraulic controls, foaming or air entrainment within the fluid can allow hunting or sponginess of the ratio control. If the drive is used where temperature variations are large, its viscosity index may be important. Traction drives do not operate successfully where the fluid viscosity is so high due to low temperature that the fluid cannot be pumped, or so thin due to high temperature than an adequate fluid film cannot be maintained.

The principal fluids now in use are derived from petroleum or are synthetic fluids compatible with petroleum formulations. The molecules of such fluids are large, and basically of three types. Those with string-type molecules are the paraffin base and are excellent lubricants for plain bearings but have relatively low traction coefficients. The traction coefficient is the ratio of traction force to contact force, as defined in Section 5.1. Lubricants with ring-type molecules have good traction coefficients but are not as good lubricants as the paraffin types and tend to oxidize and deteriorate at high temperature. These are the napthenic-base fluids. Some fluids have helical molecules and are used primarily as viscosity modifiers. The best known traction fluid is a synthetic traction fluid distributed by the Monsanto Chemical Corporation [1] under the trade name Santotrac. The comparative traction coefficients of the Santotrac fluids are about 50% higher than those of the napthenic fluids, and they are stable at high temperature.

Several companies now make or are developing synthetic traction fluids. One type under development but not yet being marketed is a silicone with a typical silicone viscosity index. The choice of fluid must be based on its intended use, and may affect a drive design. For the design of an automotive transmission that must be used in all weathers a silicone-base traction fluid with good viscosity index might be preferable even though it may have a lower traction coefficient.

Traction drives can be designed to run on any lubricating fluid. Because there is relative movement in a rolling traction contact, a nonlubricating fluid cannot be used because it cannot prevent surface interaction. A problem with traction drives is the possibility of adding the wrong fluid in the field. Many types of equipment face the same problem. The need for a special fluid should not be a deterrent.

Achievable traction coefficient is a function not only of the fluid, but of any variable that affects the thickness of the fluid film in the rolling contact area. The variables that affect the film thickness, in order of importance, are surface rolling speed, initial fluid viscosity, pressure, contact width, and contact length. The following curves present data for one traction fluid, Santotrac 50, in a way that enables appreciation of these factors.

Figure 5.11 plots the maximum traction coefficient possible at a low rolling speed as a function of mean contact pressure. Figure 5.11 is presented for pure tangential traction with no spin or other adverse factor. The curve reaches a maximum traction coefficient of 0.10 at an average contact pressure of 200,000 psi (1373 mPa), beyond which pressure the curve falls off slowly. At very low pressure the coefficient is very low as the conditions for hydrodynamic lubrication are approached.

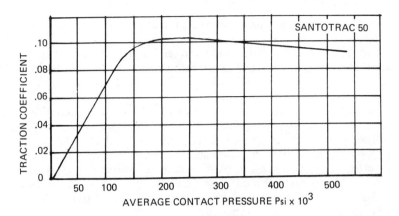

FIG. 5.11 Santotrac 50: maximum possible traction coefficient under ideal conditions.

Figure 5.12 provides a multiplier for appreciating the effect of rolling contact speed on the values of Figs. 5.11, 5.13, and 5.14. Rolling contact speed is presented in meters per second at the traction contact. Note that at higher speeds the operative traction coefficient is somewhat lower than at very low rolling contact speed.

As was shown in Section 5.1, traction force is developed only when the fluid film in the contact area is sheared, and this creates a relative motion between the traction contact members called creep. Whereas creep is a shear rate that may be expressed in inches per

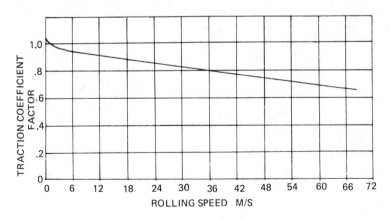

FIG. 5.12 Santotrac 50: factor to account for effect of rolling speed.

Traction Fluids

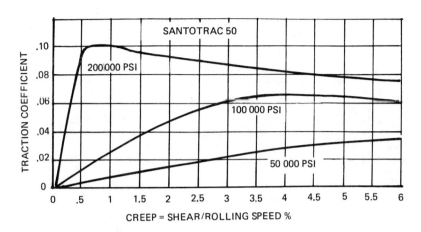

FIG. 5.13 Santotrac 50: traction coefficient as a function of creep.

second or similar units, a convenient parameter which tends to remain relatively constant is the ratio of creep distance to contact travel distance in the same time, or percent creep. Figure 5.13 presents traction coefficient as a function of percent creep for several values of mean contact pressure. From Fig. 5.13 it can be appreciated that Fig. 5.11 presented only the peak traction coefficient which could be obtained as a function of contact pressure but did not show that the availability of the maximum traction coefficient is displaced toward higher values of percent creep as the contact pressure is reduced.

Figure 5.14 presents a family of curves based on the upper curve of Fig. 5.13, drawn for a contact pressure of 200,000 psi (1373 mPa). It illustrates how the usable traction coefficient of a rolling contact is affected by geometry which imposes shear motion in the contact area due to the geometry rather than due to traction. While each element of contact area will have a traction force generated as predicted by a pure traction data curve such as the 200,000-psi curve of Fig. 5.13, the summation of usable forces in the rolling direction can be considerably reduced by the direction of the forces in the various area elements. The curves in Fig. 5.14 are not exaggerated. Some geometries have shear motions even higher than the examples used. Figure 5.14 was plotted from data available for 200,000-psi (1373-mPa) contact pressure. Drives with high geometry shear losses seldom are designed for pressures over about 140,000 psi (961 mPa) and the effect of geometry shear is proportionally less. It should not be assumed that drives with high geometry shear are not good drives. Some of the most successful drives on the market are of this type. The major effect is on rating. Such drives may have a high weight-to-power ratio.

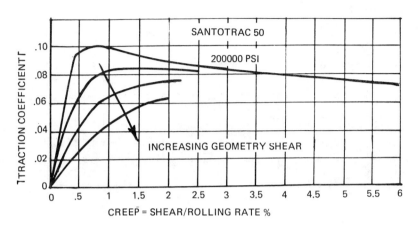

FIG. 5.14 Santotrac 50: effect of spin on traction coefficient.

The data curves of this chapter were taken from Santotrac 50 data. If a light napthenic-base hydraulic or refrigerator lubricant is used, the traction coefficient will be about 60 to 65% of the values shown. If an automatic transmission lubricant is used, the traction coefficient will be about 45% of the values for Santotrac 50, as many automatic transmission fluids have considerable paraffin base mixed in. Some drives are designed for such fluids, and manufacturers' recommendations should be followed. It should be noted that adding a higher-traction-coefficient fluid to a drive designed for a lower coefficient has very little effect, but may raise the bearing losses. There is evidence that a high-traction-coefficient fluid may raise antifriction bearing life higher than equivalent viscosity lubricating oil. Santotrac increases the fluid film thickness but also increases antifriction bearing losses.

Table 5.1 gives values of μ_{max} for 10 candidate traction fluids at the same operating condition:

Temperature, 180°F (82.2°C)
Rolling rate, 67 ft sec^{-1} (20.4 m/sec^{-1})
Mean contact pressure, 140,000 psi (965 mPa)
No spin

As pointed out in Section 5.1 and illustrated in Fig. 5.8, good design practice demands that the working range of μ not be greater than three-fourths of μ_{max} for the most adverse operating condition. The effect of operating conditions on μ is discussed in Sections 5.1, 5.2, and 5.4 and is summarized in many illustrations in Chapter 6.

Spin

TABLE 5.1 μ_{max} Values at 67 ft sec^{-1}, 180°F, and 140,000 psi

Fluid	Average μ_{max}
Polyester (Mil-L-23699)	0.035
Diester (Mil-L-7808)	0.04
Silicate esters	0.045
Polyglycols	0.045
Paraffinic	0.050
Automatic transmission fluids	0.055
Phosphate esters	0.060
Napththenic	0.058-0.065
Silicone	0.075[a]
Santotrac	0.09-0.95

[a]Some silicone fluids show decided reduction of μ_{max} after prolonged slip testing in the range 1.5 in. sec^{-1}.

5.4 SPIN

Of all the variables involved in traction design, spin is probably the most important and is the least understood. Consider an automobile rolling straight down a paved road (Fig. 5.15a). The steel-belted radial tires on the approximately 2-ft-diameter wheels roll without any relative motion between the tread and the road at any point in the contact pad. There is no surface wear and the tire could last 40,000 to 80,000 miles. Suppose that this same tire with its same load were made continuously to turn tight corners. Suppose, for example, that the turning radius was only 1 ft (Fig. 5.15b). For each revolution of the wheel, the tire would have made a 360°-turn. A rotational motion of one full turn would have been applied to the contact pad in contact with the pavement and the tread surface would wear at a high rate. This superimposed rotation is called spin and is imposed in varying degrees in all variable-ratio traction drives, and in some fixed-ratio drives. It may be, as in some small drive designs, that when one wheel is allowed to come to a stop by the drive wheel contact going to its center, spin is actually present when rotation is not present.

The analogy with a tire can be carried a step further. If the tire is turning about a very large radius, the contact pad may stay

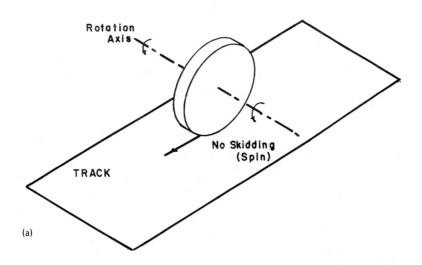

(a)

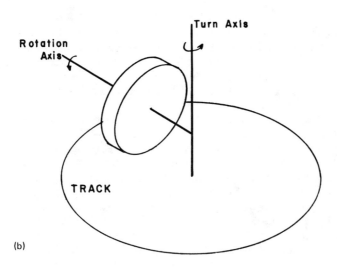

(b)

Spin

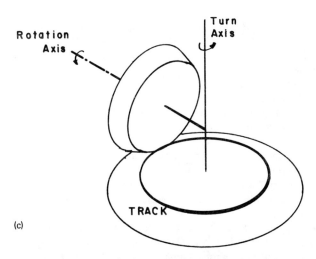

(c)

FIG. 5.15 (a) Tire on a flat track rolling straight ahead. (b) Tire in a very tight turn. Skidding (spin) occurs because the outer edge of the tire has to travel farther than the inner edge to accommodate the turn axis, yet must travel the same distance as the inner edge to accommodate the rotation axis. (c) Specially shaped tire on a specially shaped track. Skidding (spin) is reduced to zero by the special shape of tire and track. An imaginary extension of the contact line would pass through the intersection of the rotation axis and the turn axis.

in contact with the pavement without any relative motion because the tread material is elastic and deflects internally enough to allow for the relatively small amount of spin. As the turning circle becomes smaller, a rate of spin will be reached where this internal tread squirm is insufficient and scrubbing or relative motion between the tread and pavement results. In traction drives both the tire and pavement are hard steel, and internal deflection is not a practical means of taking care of any spin. However, in such drives there is a pad of very high viscosity fluid separating the two contacting surfaces. This pad transmits tangential force from the driving to the driven surface by shear resistance in the pad material. Such shear resistance is almost directly proportional to the applied force up to some point where, like the tread material, it can no longer support such a direct force-shear relationship. Excessive shear causes increased heat and loss of viscosity or loss of separation of the metal surfaces. The capacity to transfer force breaks down, the pad is ground up, and the metal surfaces contact and wear on each other.

It follows, then, that there is a maximum rate of shear permissible in a traction contact dependent on the contact force, pad material, and even to some extent on the surface finish, below which there is absolutely no metal contact and no surface wear. Spin imposes a shear rate, maximum at the outer edge of a contact area, and a uniform shear rate is imposed by any tangential force that is present. As would be expected, therefore, if tangential forces are low, a higher amount of spin can be involved before the pad shear becomes excessive. If the tangential force is high, very little spin may cause some wear to occur at the edge of a contact area. It is obvious that the amount of wear would also be a function of the amount of contact pressure. When the average contact pressure P_m is high, the pad thickness is reduced, its viscosity is high, and its ability to carry tangential force is increased. Wear in the presence of excessive shear and pad breakdown is also high. If the average contact pressure is low, the pad thickness increases and the available coefficient of traction may decrease badly. The pad, being thicker, can safely carry more shear, and when it breaks down, the wear of the metal parts may be negligible.

It has been said that P_m values above about 100,000 psi are not practical and that high-power traction drives are not possible unless they are made very large and heavy. This is not so. Rolling element bearings operate successfully at P_m values up to 400,000 psi and higher (for limited life application) and many, such as angle thrust ball bearings, have very high spin and high pad shear rates at the edges of their contact areas. If the shear rate caused by the spin is reduced by decreasing the spin, the difference can be used to support appreciable tangential forces. The very high P_m values do not automatically result in wear. A designer, by controlling the sum of the shear rates due to spin and tangential force to values within the capability of the pad material, can successfully apply P_m values approaching those used in antifriction bearing ratings.

In the example above, a wheel or traction roller was rotated on a flat pavement or disc at a radius equal to the radius of the wheel. For each revolution there was one revolution of spin. If the disc or

Spin

roller were rotated at 1800 rpm or 60 rps there would be a spin of $60 \times 2 \times \pi$ or 377 rad sec^{-1}. Suppose that both roller and disc contact surfaces were at 45° to their axes, instead of, as originally described, 0° for the roller and 90° for the disc (Fig. 5.15c). We would now find that a plane drawn tangent to the contact zone would pass through the point where the roller and disc axes intersect. Teeth could be put on both roller and disc, and a pair of miter gears would result. Spin has become zero. Between these two extreme examples, spin could be any value between 377 and 0 rad sec^{-1} for the 1800-rpm rolling speed used for illustration. Radians per second is easily understood as a measure of spin, but it is not adequate to measure the effect of spin. Spin is found to be a function not only of the radii involved but also of the angle of inclination of the contact. Since we are interested in the fluid film shear rate caused by the spin, we must also know the width of the contact area. The fluid film shear rate due to a given spin rate would be twice as great if the contact width or length were twice as great. If the contact dimensions must be known anyway, it would be convenient to express the value for spin in a way that can be used directly in analyzing what happens within a traction contact.

If spin rate and transverse contact diameter are known, the maximum rate of fluid film shear due to spin can be computed as follows:

$$\text{Film shear rate (in. sec}^{-1}) = \frac{\text{spin (rad sec}^{-1}) \times \text{TCD}}{2}$$

where TCD is the transverse contact diameter in inches. Knowing all of the geometric and load conditions of a given drive, it is possible to compute the shear rate due to spin at the outer point of the transverse contact diameter divided by the rolling speed, and to use this as a convenient measure of spin, even though this quotient changes as transverse contact diameter changes with operating conditions.

Spin thus defined can be computed as follows. Figure 5.16 shows a generalized schematic of a traction contact with its inner and outer radii labeled. R_a is the radius to the outer point of the transverse diameter of the contact. R_c is the matching point on the disc. R_b is the radius to the inner point of the transverse diameter of the contact on the roller, and R_d is the matching radius on the disc. R_a may be equal to R_b or smaller than R_b depending on the orientation of the roller to the disc.

The following relationships are apparent:

$$\frac{2 \times (R_c + R_d)}{2} = R_c + R_d$$

$$= \text{mean rolling diameter on the disc}$$

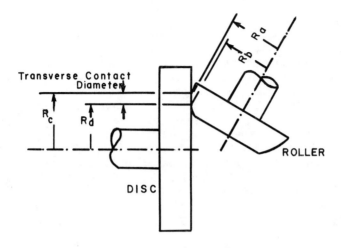

FIG. 5.16 Generalized schematic of the traction coefficient.

$$\frac{2 \times (R_a + R_b)}{2} = R_a + R_b$$

$$= \text{mean rolling diameter on the roller}$$

$$\frac{R_c + R_d}{R_a + R_b} = \text{speed ratio}$$

If the diameters are given in inches and the speed is given in revolutions per minute (rpm), then the linear velocity, called the rolling velocity, at the outer edge of the transverse contact diameter on the disc is

$$\text{Rolling velocity } VR_d = \frac{\text{rpm} \times R_c \times 2 \times \pi}{60}$$

The corresponding linear velocity at the outer edge of the transverse contact diameter on the roller is

$$\text{Rolling velocity } VR_r = \frac{(R_c + R_d) \times \text{rpm} \times R_a \times 2 \times \pi}{(R_a + R_b) \times 60}$$

The difference between these two values is the fluid film shear rate due to spin, in inches per second.

Energy Factor 165

Fluid film shear rate due to spin, $VS = VR_d - VR_r$ in. sec^{-1}

As used in this book, spin is quantified as

$$\text{Spin} = \frac{VS}{VR}$$

where

VS = fluid film shear rate due to spin at the extreme edge of the transverse contact diameter, in. sec^{-1}
VR = mean rolling rate at the center of the contact area, in. sec^{-1}
$= \dfrac{\text{rpm}_{\text{disc}} \times (R_c + R_d) \times \pi}{60}$

VR is the same for the disc and the roller, except for the effects of creep, which are not considered.

5.5 ENERGY FACTOR

When a complete traction contact analysis is done by the designer, the energy lost within the contact area is calculated. This is done by dividing the contact area into small subareas and multiplying all the small force components by the shear rate in that area. This value may be expressed in foot pounds per second, and the sum of the figures for all the small areas that comprise the total contact area is the power loss in the contact.

This is a running, or operating loss. It is a total of both spin loss and creep loss. It should be noted that it is computed with a normal fluid pad and does not apply to startup conditions. At startup, the normal pad is not yet present, and drives with high spin sometimes show excessive startup friction, particularly if the start is attempted in a speedup ratio.

If the energy loss within the contact area can be absorbed within the fluid pad by fluid shear without developing sufficient heat to drastically change pad properties, there is little effect on the traction surfaces. When the energy loss rate exceeds this amount, there is a breakdown of the fluid film and a beginning of metal contact. The amount of energy that can be absorbed per unit time is a function of the fluid pad thickness and the fluid properties under the conditions in the contact zone. It is affected therefore, by incoming fluid viscosity, contact width, and rolling rate. The fluid properties in the pad and the mean contact pressure control the shear rate which is required to support the tangential force. The amount of spin adds to the shear rate to increase the energy loss.

The rate of energy loss is not constant all over the contact area if spin is present. As a result, some strips of contact track often show surface wear or fluid discoloration effects, whereas other parts of the track are not affected. The problem of surface wear is almost entirely due to spin upsetting the uniformity of the shear rate components in the contact area and increasing local shear rate beyond safe threshold values in some areas of the contact. It follows that control of spin is the most essential factor the designer has to treat to prevent surface wear.

It is useful to compute an energy factor which is, as nearly as possible, a measure of energy dissipation in critical contact areas. For this computation, it is useful to search the contact for track strips of the highest energy loss rate, which can be done easily by modern computers. The factor as used at present by the author of this chapter is computed as follows:

$$\text{Energy factor} = \frac{\text{energy loss in strip}}{(\text{pad factor}) \times (\text{width of strip}) \times K}$$

where K is an arbitrary factor, 150,000. The pad factor is approximated by the following:

$$\text{Pad factor} = C \times V^{0.7} \times S^{0.7} \times \frac{D^{0.43}}{F^{0.13}} \times \text{fluid factor}$$

where

V = viscosity, lb sec in.$^{-2}$
S = shear rate, ft sec^{-1}
D = rolling diameter in.
F = contact force, lb
C = a factor equal to 1.05

[*Note*: lb sec in.$^{-2}$ = centipoise × 1.45 × 10^{-7} (see Section 4.2.9).] The fluid factors are:

For Santotrac 50, 0.95
For napthenic oil, 0.67
For paraffinic oil, 0.45
For Mil-L-23699, 0.33

TABLE 5.2 Relationship Between Energy Factor and Wear

Energy factor	Observed results
4	Incipient wear begins to be noted
6-8	Relatively fast wear; grooving begins
8-10	Destructive surface galling and total breakdown

The resulting energy factor, therefore, is a function of the energy loss per unit time in a strip of traction track, per unit length, and the volume of fluid passing through the contact in that same unit time, to absorb or carry away the energy. It is recognized that all the exponents involved are themselves functions of other variables. Therefore, this factor is useful only as an indication of wear danger. The method of computing it has evolved to match empirical results (see Table 5.2). Future changes to suit new data and new methods of calculation are inevitable.

Energy factor is a useful tool, but should be used only as such. Much is yet to be learned about the phenomena of excess shear and pad breakdown effects.

REFERENCE

1. Monsanto Industrial Chemicals Co., 800 N. Lindbergh Blvd., St. Louis, Mo. 63166.

6
CONTACT AREA ANALYSIS

Chapter 5 described some of the fundamental parameters of traction. Whereas it is possible, for purposes of illustration, to separate the variables and show how traction coefficient varies with contact pressure at low speed in the absence of spin (Fig. 5.8), it never happens that way. Spin is always present, rolling speed need not be low, and temperatures do vary.

Those who design traction drives have developed computer programs which break down the traction contact zone into many small strips. The shears that result from supporting the traction forces and from spin are analyzed for each small area. The analysis has to be iterative because the losses that are calculated generate heat, and heat affects the properties of the fluid in the traction contact, and these changes affect the distribution of shears in the pad. Well-written programs converge rapidly and enable the designer to examine many promising combinations of variables over a wide range of operating conditions to be certain that a safe and efficient design has been put forward. The computer also allows designers to examine a very wide range of variables so that they can see how the relationships change and can explore the most promising range.

The curves of Figs. 5.6-5.11 were intentionally left without specific ranges because they were drawn to illustrate particular trends without bringing in the complex relationships that actually govern the contact phenomena.

This chapter is compiled from *Rolling Traction Analysis and Design* by Charles E. Kraus, published in 1972 by Excelermatic, Inc., Austin, Texas.

Contact Area Analysis

The accompanying series of 26 curves (Figs. 6.1-6.26) was drawn from computer-generated data for the following case:

Temperature, 150°F (65.5°C)
Fluid, Santotrac 50
Rolling diameter, 3.4 in.
TCD/RCD, 3.44

TCD is contact area diameter transverse to rolling direction, and RCD is contact area diameter parallel to rolling directions.

Each curve includes data stating the mean contact pressure, and also the contact force and the major and minor diameters of the contact ellipse. The major diameter is TCD and the minor diameter is RCD, as noted above. Chapter 7 describes how these figures are obtained from the dimensions of the drive members and the contact force, and how the contact pressure affects life. Table 6.1 is a guide to the contact area analysis curves.

The method of testing does have a significant effect on the values of the traction coefficient, µ. In recent tests using equipment specially designed to resist even small misalignments, investigators have found that the values noted herein are conservative. Traction coefficients approximately 20% higher than those tabulated herein have been reported.

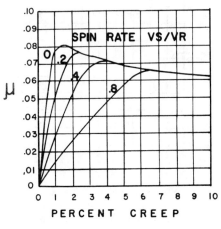

FIG. 6.1 Data: VR = 50 ft sec^{-1}, P_m = 200,000 psi, CF = 2800 lb, TCD = 0.248 in., RCD = 0.072 in.

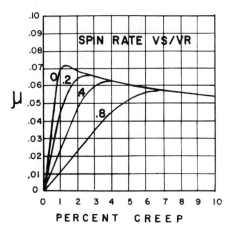

FIG. 6.2 Data: VR = 150 ft sec^{-1}, P_m = 200,000 psi, CF = 2800 lb. TCD = 0.248 in., RCD = 0.072 in.

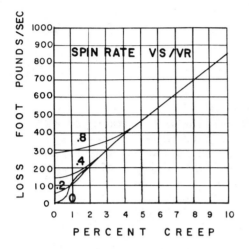

FIG. 6.3 Data: VR = 50 ft sec^{-1}, P_m = 200,000 psi, CF = 2800 lb, TCD = 0.248 in., RCD = 0.072 in.

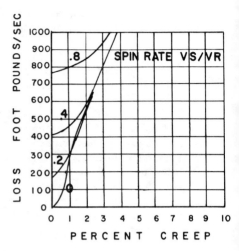

FIG. 6.4 Data: VR = 150 ft sec^{-1}, P_m = 200,000 psi, CF = 2800 lb, TCD = 0.248 in., RCD = 0.072 in.

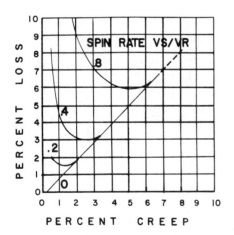

FIG. 6.5 Data: VR = 50 ft sec^{-1}, P_m = 200,000 psi, CF = 2800 lb, TCD = 0.248 in., RCD = 0.072 in.

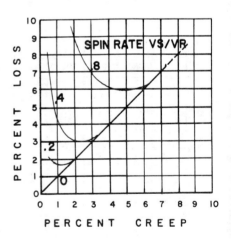

FIG. 6.6 Data: VR = 150 ft sec^{-1}, P_m = 200,000 psi, CF = 2800 lb, TCD = 0.248 in., RCD = 0.072 in.

Contact Area Analysis

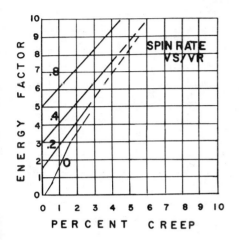

FIG. 6.7 Data: VR = 50 ft sec^{-1}, P_m = 200,000 psi, CF = 2800 lb, TCD = 0.248 in., RCD = 0.072 in.

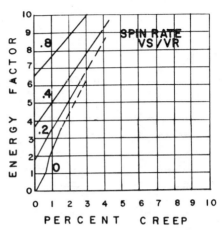

FIG. 6.8 Data: VR = 150 ft sec^{-1}, P_m = 200,000 psi, CF = 2800 lb, TCD = 0.248 in., RCD = 0.072 in.

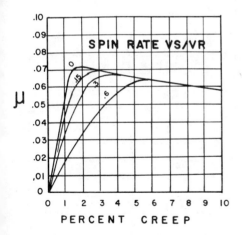

FIG. 6.9 Data: VR = 50 ft sec^{-1}, P_m = 150,000 psi, CF = 1185 lb, TCD = 0.186 in., RCD = 0.054 in.

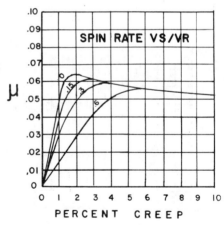

FIG. 6.10 Data: VR = 150 ft sec^{-1}, P_m = 150,000 psi, CF = 1185 lb, TCD = 0.186 in., RCD = 0.054 in.

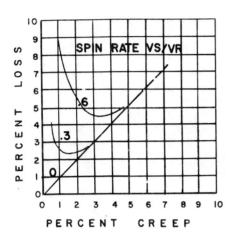

FIG. 6.11 Data: VR = any ft sec^{-1}, P_m = 150,000 psi, CF = 1185 lb, TCD = 0.186 in., RCD = 0.054 in.

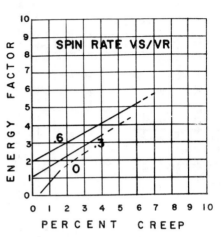

FIG. 6.12 Data: VR = 50 ft sec^{-1}, P_m = 150,000 psi, CF = 1185 lb, TCD = 0.186 in., RCD = 0.054 in.

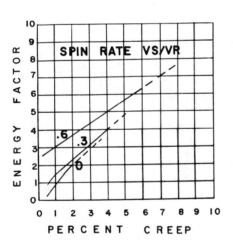

FIG. 6.13 Data: VR = 150 ft sec^{-1}, P_m = 150,000 psi, CF = 1185 lb, TCD = 0.186 in., RCD = 0.054 in.

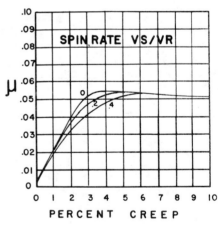

FIG. 6.14 Data: VR = 50 ft sec^{-1}, P_m = 100,000 psi, CF = 350 lb, TCD = 0.124 in., RCD = 0.036 in.

Contact Area Analysis

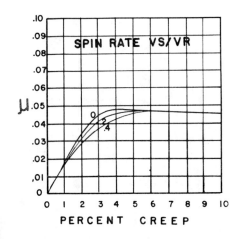

FIG. 6.15 Data: VR = 150 ft sec^{-1}, P$_m$ = 100,000 psi, CF = 350 lb, TCD = 0.124 in., RCD = 0.036 in.

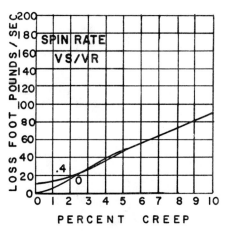

FIG. 6.16 Data: VR = 50 ft sec^{-1}, P$_m$ = 100,000 psi, CF = 350 lb, TCD = 0.124 in., RCD = 0.036 in.

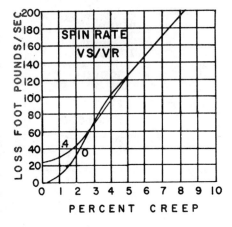

FIG. 6.17 Data: VR = 150 ft sec^{-1}, P$_m$ = 100,000 psi, CF = 350 lb, TCD = 0.124 in., RCD = 0.036 in.

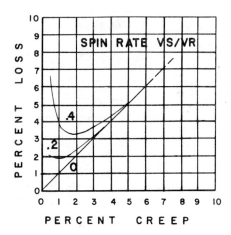

FIG. 6.18 Data: VR = any ft sec^{-1}, P$_m$ = 100,000 psi, CF = 350 lb, TCD = 0.124 in., RCD = 0.036 in.

174 Contact Area Analysis

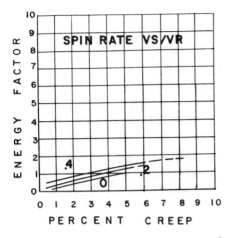

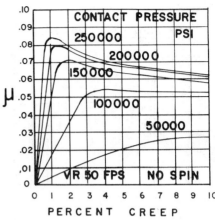

FIG. 6.19 Data: VR = 50 ft sec^{-1}, P_m = 100,000 psi, CF = 350 lb, TCD = 0.124 in., RCD = 0.036 in.

FIG. 6.20

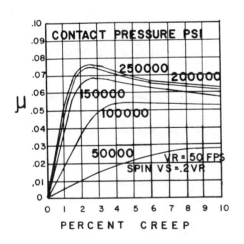

FIG. 6.21

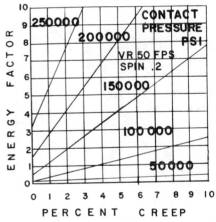

FIG. 6.22

Contact Area Analysis

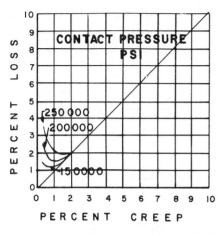

FIG. 6.23 Data: VR = 50 ft sec^{-1}, VS = 0.2VR

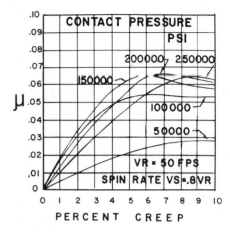

FIG. 6.24

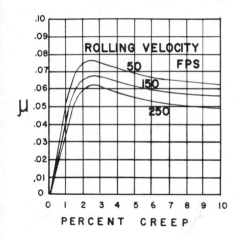

FIG. 6.25 Data: VS = 0.2VR, P_m = 200,000 psi, CF = 2800 lb, TCD = 0.248 in., RCD = 0.072 in.

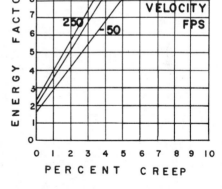

FIG. 6.26 Data: VS = 0.2VR, P_m = 200,000 psi, CF = 2800 lb, TCD = 0.248 in., RCD = 0.072 in.

TABLE 6.1 Index to Contact Area Analysis Curves

Figure	Dependent variable	Independent variable	For a family of:	Contact pressure (psi)	Roll rate (ft sec^{-1})	Spin roll ratio
6.1	μ	% Creep	Spin rates	200,000	50	
6.2	μ	% Creep	Spin rates	200,000	150	
6.3	Loss rate	% Creep	Spin rates	200,000	50	
6.4	Loss rate	% Creep	Spin rates	200,000	150	
6.5	% Loss	% Creep	Spin rates	200,000	50	
6.6	% Loss	% Creep	Spin rates	200,000	150	
6.7	Energy factor	% Creep	Spin rates	200,000	50	
6.8	Energy factor	% Creep	Spin rates	200,000	150	
6.9	μ	% Creep	Spin rates	150,000	50	
6.10	μ	% Creep	Spin rates	150,000	150	
6.11	% Loss	% Creep	Spin rates	150,000	Any	
6.12	Energy factor	% Creep	Spin rates	150,000	50	
6.13	Energy factor	% Creep	Spin rates	150,000	150	
6.14	μ	% Creep	Spin rates	100,000	50	
6.15	μ	% Creep	Spin rates	100,000	150	
6.16	Loss rate	% Creep	Spin rates	100,000	50	
6.17	Loss rate	% Creep	Spin rates	100,000	150	
6.18	% Loss	% Creep	Spin rates	100,000	Any	
6.19	Energy factor	% Creep	Spin rates	100,000	50	
6.20	μ	% Creep	Contact pressure		50	0.0
6.21	μ	% Creep	Contact pressure		50	0.2
6.22	Energy factor	% Creep	Contact pressure		50	0.2
6.23	% Loss	% Creep	Contact pressure		50	0.2
6.24	μ	% Creep	Contact pressure		50	0.8
6.25	μ	% Creep	Roll rates	200,000		0.2
6.26	Energy factor	% Creep	Roll rates	200,000		0.2

7

STRESS AND LIFE

7.1 INTRODUCTION

This chapter develops the relationships between traction drive geometry and stress for a given load. It then develops the relationships between stress and life. These relationships are statistical ones in which the probability of survival depends not only on stress but upon time as well.
    The traction drive designer uses these relationships to establish the basic rating for a particular model. The traction drive *user* need not go into all of this except to obtain an understanding of what is involved and how various types of drives may have different load/life characteristics. Given the manufacturer's rating for a given drive and user's own load spectrum, he or she may make fairly simple calculations and arrive at the probable life of the drive in the intended application.

7.2 HERTZ STRESSES

Most traction drives are based on torque transmission through conical, cylindrical, spherical, or toroidal elements which bear directly on each other with considerable pressure. The stresses induced at the contact point may be calculated by the methods described in this chapter, and from these the probability of the survival of the drive for any period of time may be estimated.
    For the case of spheres rolling on other surfaces, the stress "footprint" is elliptical. The complete theoretical calculation resorts to elliptic integrals. The practical method described below requires only that an interpolation be made in a table which has been derived for the range of the argument.

For the case of cylinders rolling on other cylinders, the stress footprint is rectangular. The calculation of the peak stress, as described, is a straightforward matter. When cones are rolled against cylinders or other cones, it is recommended that the cones be treated as cylinders of the mean contact diameter.

7.2.1 General Case

There is an excellent summary of stress and footprint formulas in Roark and Young [1]. Table 33 of this source summarizes 10 different geometries. Case 4, the general case of two bodies in contact, is reprinted here as Table 7.1.

7.2.2 Specific Cases

Sphere on Cone

Figure 7.1 represents a sphere rolling on a conical surface as might be found in a Contraves drive (Fig. 3.41), a Heynau free-ball drive (Fig. 3.43), or similar unit. The principal radii of curvature are defined on the figure.

Assume that the cone half-angle is 70°. As shown in the figure, the values chosen are:

Radius of sphere, 0.75 in.
Radius to contact point on the cone, 3.00 in.
Modulus of elasticity E for steel, 30E6 lb in.$^{-2}$
Poisson's ratio for steel, ν, 0.3
Contact force 50 lb

As shown in Fig. 7.1, the cone is a concave surface and therefore is represented as having a negative radius of curvature. Curvature is considered negative when the center is outside the material.

The contact stress, or Hertz stress, is calculated in accordance with Section 7.2.1 as follows.

Step 1. Establish radii of curvature of both bodies.
For the cone: R_1 is the curvature in the tangential direction, that is, in the direction of rotation of the cone axis.

$$R_1 = \frac{-Y}{\cos \gamma}$$

where Y is the radius to the contact point measured from the axis of rotation and γ is the half-angle of the cone. So

TABLE 7.1 Formulas for Stress and Strain Due to Pressure on or between Elastic Bodies[a]

Conditions and case no.	Formulas
4. General case of two bodies in contact: P = total load 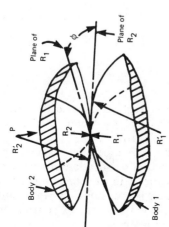	At point of contact minimum and maximum radii of curvature are R_1 and R_1' for body 1, and R_2 and R_2' for body 2. Then $1/R_1$ and $1/R_1'$ are principal curvatures of body 1, and $1/R_2$ and $1/R_2'$ of body 2; and in each body the principal curvatures are mutually perpendicular. The radii are positive if the center of curvature lies within the given body, i.e., the surface is convex, and negative otherwise. The plane containing curvature $1/R_1$ in body 1 makes with the plane containing curvature $1/R_2$ in body 2 the angle ϕ. Then: $$c = \alpha \sqrt[3]{PK_D C_E} \qquad d = \beta \sqrt[3]{PK_D C_E} \qquad \max \sigma_c = \frac{1.5P}{\pi cd}$$ $$\text{and } y = \lambda \sqrt[3]{\frac{P^2 C_E^2}{K_D}} \quad \text{where } K_D = \frac{1.5}{1/R_1 + 1/R_2 + 1/R_1' + 1/R_2'}$$ and α, β, and λ are given by the following table in which $$\cos\theta = \frac{K_D}{1.5}\sqrt{\left(\frac{1}{R_1}-\frac{1}{R_1'}\right)^2 + \left(\frac{1}{R_2}-\frac{1}{R_2'}\right)^2 + 2\left(\frac{1}{R_1}-\frac{1}{R_1'}\right)\left(\frac{1}{R_2}-\frac{1}{R_2'}\right)\cos 2\phi}$$

$\cos\theta$	0.00	0.10	0.20	0.30	0.40	0.50	0.60	0.70	0.75	0.80	0.85	0.90	0.92	0.94	0.96	0.98	0.99
α	1.000	1.070	1.150	1.242	1.351	1.486	1.661	1.905	2.072	2.292	2.600	3.093	3.396	3.824	4.508	5.937	7.774
β	1.000	0.936	0.878	0.822	0.769	0.717	0.664	0.608	0.578	0.544	0.507	0.461	0.438	0.412	0.378	0.328	0.287
λ	0.750	0.748	0.743	0.734	0.721	0.703	0.678	0.644	0.622	0.594	0.559	0.510	0.484	0.452	0.410	0.345	0.288

[a]P, total load (pounds); c, major semiaxis and d, minor semiaxis of elliptical contact area; y, relative motion of approach along the axis of loading of two points, one in each of the two contacting bodies, remote from the contact zone; ν, Poisson's ratio; E, modulus of elasticity. All dimensions are in inches, and all forces are in pounds. Subscripts 1 and 2 refer to bodies 1 and 2, respectively. To simplify expressions let
$$C_E = \frac{1-\nu_1^2}{E_1} + \frac{1-\nu_2^2}{E_2}$$

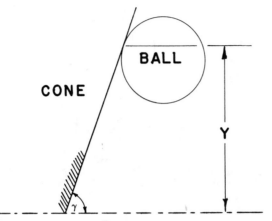

The Principal Radii of Curvature:

Cone (body 1)
  Minimum radius $R_1 = -Y/\cos \gamma$
where
  Y = radius from cone axis to point of contact
  γ = half-angle of the cone
  Maximum radius $R_1' = \infty$

Ball (body 2)
  Minimum radius $R_2 = R_{ball}$
  Maximum radius $R_2' = R_{ball}$

FIG. 7.1 Sphere on cone.

$$R_1 = \frac{-3}{\cos 70°}$$

$$= \frac{-3}{0.3420}$$

$$= -8.7714 \text{ in.}$$

R_1' is the radius of curvature at right angles to R_1. In this case it is the radius of curvature measured along a conical element, which is a straight line. Therefore, R_1' is infinite.

For the sphere: R_2 is the radius of curvature of the sphere in the same direction as the direction of rotation on the cone. That is, R_2 corresponds to R_1 except that it is measured on the sphere. Similarly, R_2' corresponds to R_1' and is measured at right angles to R_2.

In a sphere all radii of curvature are, of course, the same, and are equal to the radius of the sphere.

$R_2 = 0.75$

$R_2' = 0.75$

Step 2. Establish the curvature sum K_D. In Section 7.2.1, Roark and Young [1] defined K_D as

$$K_D = \frac{1.5}{1/R_1 + 1/R_2 + 1/R_1' + 1/R_2'} \qquad (7.1)$$

$$= \frac{1.5}{1/-8.7714 + 1/0.75 + 1/\infty + 1/0.75}$$

$$= 0.5876$$

Step 3. Evaluate the elliptic integral from Table 7.1. ϕ is the angle that the plane containing the curvature $1/R_1$ in body 1 makes with the plane containing the curvature $1/R_2$ in body 2. In this case body 2 is a sphere and it does not matter. The expression

$$\cos\theta = \frac{K_D}{1.5}\sqrt{\left(\frac{1}{R_1}-\frac{1}{R_1'}\right)^2 + \left(\frac{1}{R_2}-\frac{1}{R_2'}\right)^2 + 2\left(\frac{1}{R_1}-\frac{1}{R_1'}\right)\left(\frac{1}{R_2}-\frac{1}{R_2'}\right)\cos 2\phi}$$

has only one nonzero term under the radical because R_2 equals R_2' for the sphere.

$$\cos\theta = \frac{0.5876}{1.5}\sqrt{\left(\frac{1}{-8.7714}-\frac{1}{\infty}\right)^2}$$

$$= 0.04466$$

Interpolate in Table 7.1 to obtain

$\alpha = 1.031$

$\beta = 0.9714$

Step 4. Establish C_E, which is a factor dependent on the materials of each body. In this case both bodies are the same material. Using the values tabulated in the beginning of this section, we have

$$C_E = \frac{1-\nu_1^2}{E_1} + \frac{1-\nu_2^2}{E_2}$$

$$= \frac{1-0.3^2}{30E6} + \frac{1-0.3^2}{30E6}$$

$$= 6.0667E\text{-}8$$

Step 5. Establish the dimensions of the elliptical footprint created when the two steel bodies contact each other and deflect in the process.

$$c = \alpha(PK_D C_E)^{1/3}$$
$$= 1.031(50 \times 0.5876 \times 6.0667E\text{-}8)^{0.33333}$$
$$= 0.0125 \text{ in.}$$

$$d = \beta(PK_D C_E)^{1/3}$$
$$= 0.9714(50 \times 0.5876 \times 6.0667E\text{-}8)^{0.33333}$$
$$= 0.01178 \text{ in.}$$

The reader will recognize d as the contact diameter in the direction of rolling, referred to as RCD in Chapters 5 and 6; c will be recognized as the contact diameter transverse to the direction of rolling, referred to as TCD in Chapters 5 and 6.

Step 6. Calculate the maximum contact stress, or Hertz stress.

$$S_{c\ max} = \frac{1.5P}{\pi cd}$$
$$= \frac{1.5 \times 50}{3.1416 \times 0.0125 \times 0.01178}$$
$$= 162,147 \text{ lb in.}^{-2}$$

The maximum Hertz stress is keyed to the performance of the traction drive as described in Chapters 5 and 6. It is related to the maximum shear stress used in the Lundberg-Palmgren analysis in Section 7.3.1.

Cone-on-Cone, Cylinder-on-Cone, Cylinder-on-Cylinder

The method of the preceding case is not applicable to the cone-on-cone or cone-on-cylinder or cylinder-on-cylinder cases because these cases all have a rectangular footprint and the preceding method was developed for elliptical footprint cases only.

In cone-on-cylinder drives, slight scrubbing will take place because all the points of the cylinder are moving at one speed, but all the points on the contact line on the cone are moving at different speeds. The contact line is usually made short for this reason. When the contact line is short, very little error is encountered if both elements are treated as cylinders. The cone is taken as a cylinder of diameter equal to the mean pitch diameter.

Reference 2 develops a formula for the maximum compressive stress developed when a cylinder runs on a cylinder. Using the same notation as that used in the preceding case, and using the same values for E and ν, this formula would read

Hertz Stresses

$$S_{c\,max} = 2717\left[\frac{P}{L}\left(\frac{1}{R_1} + \frac{1}{R_2}\right)\right]^{1/2} \tag{7.2}$$

where L is the length of the contact line. The same formula may be developed from Ref. 1, cases 2b and 2c of Table 33.

Figure 7.2 is a schematic representation of a steel cone running against a steel ring, as would be the case in a Graham drive (Fig. 3.21) or a Shimpo drive (Fig. 3.24 or Fig. 3.29).

For purposes of this case consider that the cone is pressed against the ring with a force of 600 lb. Assume that the ring has a radius of 4 in. and the cone is engaged at a point along its axis where the radius of the cone is 1 in. Assume that the contact band is 0.125 in. wide.

$$S_{c\,max} = 2717\left[\frac{P}{L}\left(\frac{1}{R_1} + \frac{1}{R_2}\right)\right]^{1/2}$$

$$= 2717\left[\frac{600}{0.125}\left(\frac{1}{1} + \frac{1}{4}\right)\right]^{1/2}$$

$$= 163,000\ \text{lb in.}^{-2}$$

Sphere-on-Toroid

Figure 7.3 represents a toroidal traction drive element such as that discussed in Section 3.1.6. Figure 7.3 does not actually describe any of the four units listed in that section; it is perhaps closest to

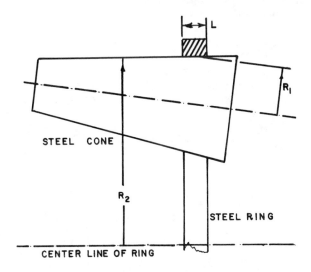

FIG. 7.2 Cylinder-on-cone.

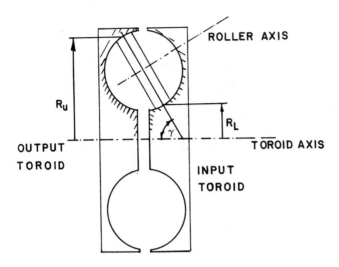

FIG. 7.3 Sphere-on-toroid.

the Kopp ball variator of Section 3.1.1. The principle of analysis is, of course, the same for all such units.

The method used in this section is the same as that used for the sphere-on-cone case except that the radii of curvature are quite different. Although both inner and outer contacts are examined, the stress is expected to be higher at the inner contact because there the torus is convex and the footprint of the sphere will necessarily be smaller.

In the following example, the longitudinal radius of the torus has been made 104% of the radius of the sphere. If the two radii were exactly the same, there would be severe skidding at the contact line. This slight difference in radius requires a lateral position adjustment mechanism to maintain the appropriate force over the angular range of the ball spin axis.

EXAMPLE. Assume a toroidal traction drive (Fig. 7.3) operating as follows:

>Radius to center of torroidal cup, 1.25 in.
>Toroid minimum radius, R_1, 0.75 in.
>Roller angle γ, 60°
>Radius to upper contact point, $1.25 + 0.75 \sin \gamma$, 1.8995 in.
>Radius to lower contact point, $1.25 - 0.75 \sin \gamma$, 0.6005 in.
>Roller spherical radius R_2, R_2', 0.721 in.
>Force between roller and torus, 50 lb

Hertz Stresses

	For outer contact point	For inner contact point

Step 1:

Minimum curvature $\quad R_1 = -0.75 \quad\quad\quad\quad\quad\quad\quad R_1 = -0.75$

Maximum curvature $\quad R_1' = \dfrac{-1.8995}{\sin \gamma} \quad\quad\quad R_1' = \dfrac{+0.6005}{\sin \gamma}$

$\quad\quad\quad\quad\quad\quad\quad\quad\quad\quad = -2.1934 \quad\quad\quad\quad\quad\quad = +0.6934$

Step 2:

$$K_D = \dfrac{1.5}{\dfrac{1}{R_1} + \dfrac{1}{R_2} + \dfrac{1}{R_1'} + \dfrac{1}{R_2'}}$$

$= \dfrac{1.5}{\dfrac{1}{-0.75} + \dfrac{1}{0.721} + \dfrac{1}{-2.1934} + \dfrac{1}{0.721}} \quad\quad \dfrac{1.5}{\dfrac{1}{-0.75} + \dfrac{1}{0.721} + \dfrac{1}{0.6934} + \dfrac{1}{0.721}}$

$= 1.52335 \quad\quad\quad\quad\quad\quad\quad\quad\quad\quad\quad\quad 0.5203$

Step 3:

$$\cos \theta = \dfrac{K_D}{1.5} \left[\left(\dfrac{1}{R_1} - \dfrac{1}{R_1'} \right)^2 \right]^{1/2}$$

Trivial terms omitted:

$= \dfrac{1.52335}{1.5} \left[\left(\dfrac{1}{-0.75} + \dfrac{1}{-2.1934} \right)^2 \right]^{1/2} \quad\quad \dfrac{0.5203}{1.5} \left[\left(\dfrac{1}{-0.75} - \dfrac{1}{0.6394} \right)^2 \right]^{1/2}$

$= .89107 \quad\quad\quad\quad\quad\quad\quad\quad\quad\quad\quad\quad\quad 0.9627$

From Table 7.1, by interpolation,

$\alpha = 3.005 \quad\quad\quad\quad\quad\quad\quad\quad\quad\quad \alpha = 4.701$

$\beta = 0.469 \quad\quad\quad\quad\quad\quad\quad\quad\quad\quad \beta = 0.371$

Step 4:

$C_E = \dfrac{1 - \nu_1^2}{E_1} + \dfrac{1 - \nu_2^2}{E_2}$

$\quad\quad = 6.0667E\text{-}8 \text{ from first example} \quad\quad 6.0667E\text{-}8$

Step 5:

$c = \alpha (P K_D C_E)^{1/3}$

$\quad = 3.005(50 \times 1.52335 \times 6.067E\text{-}8)^{1/3} \quad\quad 4.701(50 \times 0.5203 \times 6.067E\text{-}8)^{1/3}$

$\quad = 0.05005 \quad\quad\quad\quad\quad\quad\quad\quad\quad\quad\quad 0.05473$

| For outer contact point | For inner contact point |

Step 6:

$d = \beta(PK_D C_E)^{1/3}$

$ = 0.469(50 \times 1.52335 \times 6.067\text{E-}8)^{1/3}$ $0.371(50 \times 0.5203 \times 6.067\text{E-}8)^{1/3}$

$ = 0.00781$ 0.00432

Step 7:

$$S_{c\ max} = \frac{1.5P}{\pi cd}$$

$$= \frac{1.5 \times 50}{3.14 \times 0.05005 \times 0.00781} \qquad \frac{1.5 \times 50}{3.14 \times 0.05473 \times 0.00432}$$

$= 61{,}704\ \text{psi}$ $100{,}982\ \text{psi}$

7.3 LIFE

The stresses in bodies under direct bearing may go surprisingly high—higher than the published ultimate tensile strength—without causing immediate apparent damage. This may be due to several factors. Contrary to the tensile case, where parts are free to separate once a crack begins to propagate, and where the stress concentration at the tip of such a crack may generate even higher stresses, parts stressed in bearing may have no place to move. Metal stressed above the yield point may be held in place by friction or by the matrix of less highly stressed metal surrounding it. When the load is removed, the stress is relieved without apparent damage. If the process is repeated many times, failure may become evident in the form of surface or subsurface spalling (e.g., fatigue).

When the parts are bathed in lubricant and the contact between the bodies is maintained through an elastohydrodynamic film, the nature of the footprint may change sufficiently that the subsurface shear peak stresses anticipated by the elastic materials equations may not be reached. Reference 3 describes a lubrication factor, F, which is a life multiplier based on the properties of the lubricant film and the metal surfaces.

7.3.1 Life-Prediction Method for Designers

In the high-contact-stress regime, which is the only practical regime in which to operate most traction drives, the achievement of a given number of stress repetitions without failure can only be assigned a probability. Most drives are rated against the 90% probability of

surviving the number of stress repetitions associated with the rated
number of hours of operation at the rated speed and load. This is
called the L_{10} life.

In this section a method of doing this will be described. The
Lundberg-Palmgren theory has a strong basis in theory and has been
used quite successfully to predict the life of ball bearings. Many
traction drives have elements that resemble parts of large ball bearings,
so the transfer between disciplines is simple and natural. The
Lundberg-Palmgren theory has some empirical constants built into it
which related to material properties of the best ball bearing steels of
the late 1940s. Correction factors have been published [3] which
bring the results calculated from the original papers [4] into line
with what can be obtained with the best vacuum melted steels.

The calculation produces a predicted life for 90% survival. A
method is detailed whereby if the life for 90% survival is known, the
life for any other percentage survival is calculable with equal validity.
This method introduces the Weibull distribution, which is the basis
for almost all fatigue-related mathematics.

The reader who does not intend to design his or her own drive
need not pursue this section except to appreciate the design considerations
that form the basis for the rating of any drive selected or applied.
The traction drive user is directed to Section 7.3.4, in which, once
the rating for a drive has been established by its manufacturer, its
life under any anticipated set of operating conditions may be derived
simply.

The Lundberg-Palmgren Method

Once a failure mechanism can be postulated, methods can be devised
which might predict the occurrence of failure with considerable validity.
The Lundberg-Palmgren theory [4] is based on the premise that
fatigue failures begin at flaw nuclei under the surface and progress
to the surface. The stress component which is critical to this process
is the maximum alternating shear stress. Fatigue is a statistical
mechanism which depends in part on the probability of finding a
fatigue nucleus; therefore, it is dependent on the total volume of
material subject to the alternating stress.

Rohn et al. [5] describe a technique for applying the Lundberg-
Palmgren theory to traction drives. The process begins with the
calculation of the semimajor and semiminor diameters of the contact
ellipse, as has been demonstrated in Section 7.2, and the associated
peak Hertz contact stress, as has been demonstrated in the same
section.

The life calculation takes the form

$$L = K_4 K_2^{0.9} Q^{-3} \rho^{-6.3} R^{-0.9} \tag{7.3}$$

where

L = life, millions of revolutions

$$K_2 = \left(\frac{z_0}{b}\right)^{4/3} \left(\frac{\tau_0}{\sigma_0}\right)^{-31/3} (a^*)^{28/3} (b^*)^{35/3} \qquad (7.4)$$

where $\frac{z_0}{b}$ = ratio of the depth to the maximum orthogonal shear stress to the semiminor diameter of the contact ellipse; the value of this factor is read from Fig. 7.4

$\frac{\tau_0}{\sigma_0}$ = ratio of the value of the maximum orthogonal shear stress to the peak contact stress; the value of this factor is read from Fig. 7.4

$$a^* = \frac{a}{0.0045 \sqrt[3]{Q/\rho}}$$

where Q = contact force, lb (notation P was used in the reference to Roark and Young [1] in the earlier part of this chapter)

ρ = inverse curvature sum (ρ is equal to $1.5/K_D$ as used in the reference to Roark and Young [1] in the earlier part of this chapter)

a = semimajor diameter of the contact ellipse (notation c was used for this factor in the reference to Roark and Young [1] in the earlier part of this chapter)

$$b^* = \frac{b}{0.0045 \sqrt[3]{Q/\rho}}$$

where b = semiminor diameter of the contact ellipse (the notation d was used for this factor in the reference to Roark and Young [1] in the earlier part of this chapter)

K_4 = 6.43E8 for ball-bearing steel AISI 52100 heat treated to hardness RC 62

Q = contact force as defined within the description of K_2

ρ = inverse curvature sum as defined within the description of K_2

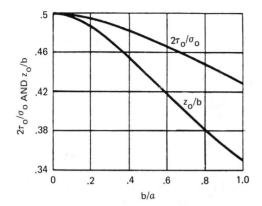

FIG. 7.4 $2\tau_0/\sigma_0$ and z_0/b versus b/a. (Courtesy of American Society of Mechanical Engineers, New York.)

R = rolling radius of the contact path;
$R \times 2 \times \pi$ determines the circumference of the path and is a factor, together with z_0 and b, in determining the volume of material subject to shear stress

Using the values for the sphere-on-cone case illustrated in Section 7.2.2, we have

$a = 0.0125$ in.

$b = 0.01178$ in.

$\rho = \dfrac{1.5}{0.5875}$

$= 2.5532$

R = 3 for the cone

Q = 50

$a^* = \dfrac{a}{0.0045 \sqrt[3]{Q/\rho}}$

$= \dfrac{0.0125}{0.0045 \sqrt[3]{50/2.5532}}$

$= 1.0305$

$$b^* = \frac{b}{0.0045 \sqrt[3]{Q/\rho}}$$

$$= \frac{0.01178}{0.0045 \sqrt[3]{50/2.5532}}$$

$$= 0.97119$$

$$\frac{b}{a} = \frac{0.01178}{0.0125}$$

$$= 0.9424$$

Enter Fig. 7.4 at $b/a = 0.9424$ and read

$$\frac{z_0}{b} = 0.355$$

$$\frac{2\tau_0}{\sigma_0} = 0.425$$

Therefore,

$$\frac{\tau_0}{\sigma_0} = 0.2125$$

$$K_2 = (0.355)^{4/3}(0.2125)^{-31/3}(1.0305)^{28/3}(0.97119)^{35/3}$$

$$= 0.2513 \times 8{,}925{,}331 \times 1.3237 \times 0.71102$$

$$= 2{,}111{,}000$$

$$L = K_4 K_2^{0.9} Q^{-3} \rho^{-6.3} R^{-0.9}$$

$$= 6.43\text{E}8 \times 2.111\text{E}6^{0.9} \times 50^{-3} \times 2.5532^{-6.3} \times 3^{-0.3}$$

$$= 6.43\text{E}8 \times 492{,}084 \times 8\text{E-}6 \times 2.725\text{E-}3 \times 0.372$$

$$= 2.566\text{E}6$$

This life is expressed in millions of cycles. Clearly, the contact load chosen to make the illustration was very light!

Every contact in the drive is analyzed in this manner, and the analysis is reduced to hours for each part. If the cone we had just analyzed were turning at 1750 rpm and there were four balls running on it, equally loaded, the contact life would be

$$L = \frac{10 \times 2.566\text{E}6}{1750 \times 60 \times 4}$$

$$= 6.11\text{E}6$$

Life

Reference 5 points out that the life of the drive can be estimated by summing up all the lives of the individual elements in accordance with the following relationship:

$$H_{sum} = \left[\frac{1}{(H_1)^e} + \frac{1}{(H_2)^e} + \frac{1}{(H_3)^e} + \cdots + \frac{1}{(H_n)^e}\right]^{-1/e} \quad (7.5)$$

where $H_1, H_2, H_3, \cdots, H_n$ are the calculated lives in hours of all the elements of the drive, and e is 10/9.

If the drive analyzed above consisted of four spheres and two cones, the life of each sphere would be calculated using the same factors except for R.

$$L = K_4 K_2^{0.9} Q^{-3} \rho^{-6.3} R^{-0.9}$$
$$= 6.43E8 \times 492,084 \times 8E\text{-}6 \times 2.725E\text{-}3 \times R^{-0.9}$$
$$= 6.898E6 \times R^{-0.9}$$

In Fig. 7.1 no axis was shown for the sphere. Suppose, as in Fig. 7.5, that the sphere axis were chosen 10.314° tilted from the cone axis so that the contact radius at the input side was 0.3786 in. and the contact radius, assuming that there were a symmetrical cone contacting the other side of the sphere, at the output side was 0.1262 in. Such an arrangement would yield a speed reduction of 3:1 from input to output.

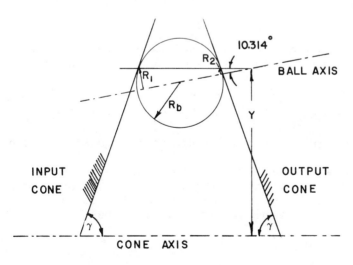

FIG. 7.5 Traction drive set up for 3:1 ratio with sphere on cones ($R_1 = 0.3786$, $R_2 = 0.1262$, $R_b = 0.750$, $\gamma = 70°$).

Life of sphere, input side:

$$L = 6.898\text{E}6 \times 0.3786^{-0.9}$$

$$= 16.533\text{E}6 \text{ millions of cycles}$$

Life of sphere, output side:

$$L = 6.898\text{E}6 \times 0.1262^{-0.9}$$

$$= 44.438\text{E}6 \text{ millions of cycles}$$

Converting to hours, note that the sphere is turning at $3/0.3786 \times$ input speed, or

$$\frac{3 \times 1750}{0.3786} = 13,866 \text{ rpm}$$

The output shaft is turning at $1750/3 = 583$ rpm.

Element lives, in hours are:

$$\text{Sphere, at input contact} = \frac{16.533\text{E}6 \times 10^6}{13,866 \times 60}$$

$$= 19.872\text{E}6 \text{ hr}$$

$$\text{Sphere, at output contact} = \frac{44.438\text{E}6 \times 10}{13,866 \times 60}$$

$$= 53.41\text{E}6 \text{ hr}$$

Output shaft life is three times input shaft life because it is running at one-third the speed, but has the same contact stress and material parameters.

$$L_{sum} = \left(\frac{1}{6.11\text{E}6^{10/9}} + \frac{4}{19.87\text{E}6^{10/9}} + \frac{4}{53.41\text{E}6^{10/9}} + \frac{1}{18.33\text{E}6^{10/9}} \right)^{-9/10}$$

$$= 2.47 \text{ million hours for the drive at this operating condition}$$

Reference 5 points out that the load Q appears in the equations to the third power. This drive could carry 10 times the load we have analyzed it for, and it would still have a fatigue life of 2470 hr. A reasonable rating, to obtain 10,000 hr of life, would be

$$Q = 50 \times \left(\frac{2.47\text{E}6}{10,000} \right)^{1/3}$$

$$= 314 \text{ lb per contact}$$

Life 193

If the traction coefficient were 0.06, the tangential force at the input radius of 3 in. would be 0.06 × 314 or 18.8 lb per contact, and the drive horsepower rating would be

$$hp = \frac{4 \times 18.8 \times 3 \times 1750}{63,025}$$

$$= 6.26 \text{ hp}$$

7.3.2 Life as a Function of Load Spectrum

The Lundberg-Palmgren method described in the preceding section was shown to predict the life of a drive as a complex function of contact force and geometry. The drive does not always operate at the same contact force at all operating conditions.

If operating conditions vary predictably, that is, if the drive operates at part load a certain percentage of the time, and if contact force is reduced in a known way as a function of load reduction, then several methods can be used to obtain an assessment of the probable life of the drive under the operating spectrum.

The Lundberg-Palmgren method could be used to predict the life under each known set of operating conditions. Then if each operating condition could be expressed as a percentage of total operating time, a composite life could be obtained:

$$\text{Life}_1 \times \%_1 + \text{life}_2 \times \%_2 + \cdots + \text{life}_n \times \%_n = \text{life}_{\text{total}}$$

This would be a tedious process, but would achieve an accurate result.

Traction drives are often designed with only moderate preload between the rolling elements. A loading device is provided to increase the contact pressure proportional to output torque.

The inverse life exponent is so large that for cases in which there are periods of less than maximum load, life is increased substantially compared to designs that run at fixed contact force (proportional to maximum rated load) at all times. One type of drive (Fig. 3.20) relies on centrifugal force to maintain contact pressure. Such a drive cannot take advantage of the load/life sensitivity relationship if it runs at partial load but at maximum speed. For those drives which have loading means proportional to output torque, and which are operated at a spectrum of output torques, a mean load can be calculated which will yield a correct mean life estimate.

The loading spectrum must be divided into periods and the contact force calculated for each. Then each force must be raised to the third power. Each value must be weighted in accordance with its percentage of total operating time. The mean force is obtained by taking the one-third power of the sum of the weighted numbers.

The Lundberg-Palmgren method need then be used but once to calculate the mean life based on the mean contact force.

EXAMPLE. Suppose that the drive of Sections 7.2.2 and 7.3.1 were analyzed for the following spectrum of operating loads:

Operating load, percent of rating	Percent of total operating time at this load
100	20
80	40
30	30
10	10
	100

A table would be constructed as follows:

Step description (% load)	Time percent at this step	Contact force, Q (lb)	Q^3	Q^3 × time percent
100	20	314	30.96E6	6.19E6
80	40	251.2	15.85E6	6.34E6
30	30	94.2	83.59E4	2.51E5
10	10	31.4	30.96E3	3.10E3
			Total of Q^3 × time percent =	1.28E7

Mean contact force = $1.28E7^{1/3}$

= 234 lb

If the life of the traction drive were analyzed by the same Lundberg-Palmgren method that yielded a life estimate of 10,000 hr for a mean contact force of 314 lb, for the mean contact force of 234 lb corresponding to the load spectrum above, the life would be found to be

$$L = 10{,}000 \text{ hr} \times \left(\frac{314}{234}\right)^3$$

$$= 24{,}163 \text{ hr}$$

Life

Note the dramatic increase in life. It is for this reason that most traction drives are preloaded by mechanisms that maintain contact forces just sufficient to carry the output torque. Drives without such mechanisms must be rated to carry the full output torque at all times. They could be simpler drives but they would have shorter lives than self-loading drives, but only when the drive is working a variable load spectrum. If the drive must work at full-rated torque at all times, both types would have the same life potential.

This analysis shows that overloading a drive would reduce its life very dramatically. A fixed preload drive could not be overloaded (beyond its preload limit) at all. It might skid and destroy itself. A self-loading drive would cope with the extra load, but with a reduction in life.

EXAMPLE. Consider the same unit analyzed above, but with this new load spectrum:

Fraction of rated load (%)	Time spent at this fraction (%)
125	10
100	20
80	30
30	30
10	10

The load spectrum table would look like this:

Step description (% load)	Time percent this step	Contact force, Q	Q^3	Q^3 × time percent
125	10	392.5	6.046E7	6.046E6
100	20	314	3.096E7	6.192E6
80	30	251.2	1.585E7	4.755E6
30	30	94.2	8.359E5	2.508E5
10	10	31.4	3.096E4	3.096E3

Total of Q^3 × time percent = 1.725E7

Mean contact force = $1.725E7^{1/3}$

= 258.4 lb

$$\text{Life} = 10{,}000 \times \left(\frac{314}{258.4}\right)^3$$

$$= 17{,}950 \text{ hr}$$

In the two load spectrum examples the time at 100%, 30% and 10% load was the same. The difference was that instead of 80% load 40% of the time, we made provision for 80% load 30% of the time and added 125% load for 10% of the time. The predicted life was reduced to 75% of its former value by the change.

7.3.3 Relationship Between Time and Percent Survival

The life-prediction technique described in Sections 7.3.1 and 7.3.2 enabled the designer or the user to predict the life at which either 1% or 10% of the devices might be expected to fail (e.g., the life that would be exceeded by 99% or 90% of the population, respectively). The life-prediction process is not restricted to examining these particular end points only. This section describes the techniques to be used to examine the life that any desired fraction of the population will probably achieve.

Most statistical studies are based on the Poisson (constant failure rate) distribution, or on a slightly more complex relationship called the Weibull distribution.

Poisson Distribution

This is the simplest form of statistical distribution for it is based on constant failure rate.

$$P_S = e^{-(\lambda t)}$$

where

P_S = probability of survival
 = 1 - L, where L is the percent failed
e = base of natural logarithms
 = 2.71828 . . .
λ = failure rate per unit time
t = number of time units being examined

λ is also sometimes expressed as k/d, where d is the L_{10} life and k is a factor calculated to establish the initial boundary of the equation. Suppose that the L_{10} life of a particular drive under a particular set of operating conditions is 6000 hr and the operating time under those conditions is 6000 hr. Then the L_{10} life will have been reached and

Life

the P_S would be expected to be 0.90. The boundary would be established as follows:

$$P_S = e^{-[k(6000/6000)]}$$
$$0.9 = e^{-(k)}$$

Taking logarithms to the base e of both sides, we have

$$-0.10536 = -k$$
$$k = 0.10536$$
$$P_S = e^{-[0.10536(t/d)]}$$

If this distribution were valid for fatigue analyses, the probability of survival of any system could be predicted by substituting the fraction of L_{10} life that would be of interest. Unfortunately, the Poisson distribution does not track the failure rate of metals stressed in fatigue. For example, using the Poisson distribution in an attempt to predict L_{50}, we have

$$P_S = e^{-[k(t/d)]}$$
$$0.5 = e^{-[k(t/d)]}$$

Taking logarithms of both sides to the base e yields

$$-0.6931 = -k\frac{t}{d}$$

From the preceding paragraph the boundary condition was found to be $k = 0.10536$, so

$$\frac{t}{d} = \frac{0.6931}{0.10536}$$
$$= 6.57$$

Analysis of fatigue test data by Lieblein and Zelen [6] shows that the t/d ratio for 50% failures is much shorter than that. A statistically reasonable value for t/d at L_{50} is 4.08, or at the most, 5.0. K is not a constant in the mathematics of fatigue, and resort must be made to the Weibull distribution to provide life predictions that correlate well with experimental results.

Weibull Distribution

The boundary conditions of the Weibull distribution have to be established at two points because this distribution has a second boundary constant, b.

$$P_S = e^{-[k(t/d)]^b}$$

where

P_S = probability of survival
  = 1 - L
e = base of natural logarithms
  = 2.71828 ...
$\frac{t}{d}$ = ratio of operating life at the point where P_S is being evaluated to the design life L_{10}, where $P_S = 0.90$
k = boundary constant
b = boundary constant

The exponent b and the constant k are evaluated from test data by solving the equation above for test lives at 10% and 50% failure simultaneously. Lieblein and Zelen [6] determined b to be 1.32497 in a study of the fatigue life of deep-groove ball bearings. Reference 7 presents a solution based on the ratio of L_{50} to L_{10} of 4.08, for which case b would be 1.34.

If one accepts the value of b as 1.32497, the value of k is easily established. At the L_{10} life, t/d is 1.0, so

$$0.9 = e^{-(k)^{1.32497}}$$

Taking the natural logarithms of both sides of the equation, we have

$$-0.10536 = -(k)^{1.32497}$$

Clearing the minus signs and taking the logarithms again yields

$$-2.25036 = 1.32497(\log_e k)$$
$$\log_e k = -1.6984$$
$$k = 0.18297$$

Using the equation thus found to evaluate the ratio of any life L_{xx} to life L_{10} is a simple task.

$$P_S = e^{-[0.18297(t/d)]^{1.32497}}$$

EXAMPLE. Evaluate t/d for L_{50}.

$$0.50 = e^{-[1.8297(t/d)]^{1.32497}}$$

Taking the natural logarithms of both sides of the equation, we have

$$-0.693147 = -\left(0.18297 \frac{t}{d}\right)^{1.32497}$$

Clearing the minus signs and taking the logarithms again yields

$$-0.36651 = 1.32497 \left(\log_e 0.18297 \frac{t}{d}\right)$$

$$-0.276619 = \log_e 0.18297 \frac{t}{d}$$

Taking the antilog$_e$ of both sides, we obtain

$$0.75834 = 0.18297 \frac{t}{d}$$

$$\frac{t}{d} = 4.14$$

Table 7.2 is a summary of the ratio of operating life to L_{10} design life for a wide range of the argument P_s. If plotted on log-log paper, the information in the table would not present a straight line. Special Weibull paper (Fig. 7.6) is available on which Weibull distributions plot as straight lines. The paper is available from Technical Engineering Aids for Management (104 Belrose Avenue, Lowell, MA 01065).

7.3.4 User's Method

General Case: Ball-on-Cone Drive or Toroidal Drive

The user need not know the actual stresses or contact forces inside the drive to obtain an estimate of its life under an intended spectrum of operating conditions. First, obtain the manufacturer's rated life for operation 100% of the time at rated load. Then divide the spectrum of intended usage into steps expressed as a fraction of rated load. The influence of each step of intended usage is inversely proportional to the cube of the load ratio at that step times the percentage of time the drive is to be used at that step. The sum of all of these is the inverse ratio of drive life under this spectrum of loading compared to the manufacturer's rated life for continuous 100% loading.

TABLE 7.2 Ratio of Operating Life to L_{10} Life for Various Probabilities of Survival

Probability of survival for operating life	L value	Ratio of allowable operating life to L_{10} life
0.995	0.5	0.1030
0.99	1	0.1785
0.98	2	0.2910
0.97	3	0.396
0.96	4	0.492
0.95	5	0.584
0.94	6	0.672
0.93	7	0.756
0.92	8	0.840
0.91	9	0.921
0.90	10	1.000
0.85	15	1.383
0.80	20	1.750
0.75	25	2.120
0.70	30	2.435
0.65	35	2.860
0.60	40	3.250
0.55	45	3.600
0.50	50	4.080
0.45	55	4.540
0.40	60	5.03
0.35	65	5.56
0.30	70	6.16
0.25	75	6.84
0.20	80	7.65
0.15	85	8.64
0.10	90	10.00
0.05	95	12.40
0.01	99	16.40

Source: Ref. 7.

Life

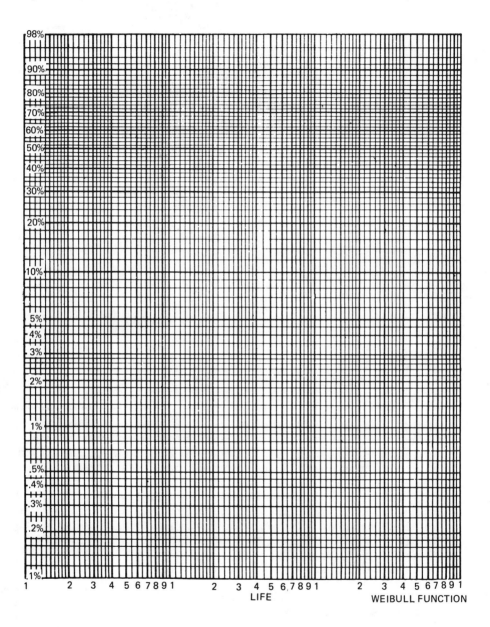

FIG. 7.6 Weibull function paper.

EXAMPLE. Suppose that the drive being examined in Section 7.3.2 were being evaluated by a user who did not know the internal stresses in the drive. The user knows only that the manufacturer has rated this drive for a 10,000 hr life at full load. We intend to use it at the spectrum of loads of the second example of that section.

Construct the following table:

Step description (load fraction)	Time percent at that step	Contact life inverse ratio (step descr.[3])	Contact life inverse ratio × time percent at that step
1.25	10	1.953	0.1953
1.00	20	1.000	0.2000
0.80	30	0.512	0.1536
0.30	30	0.027	0.0081
0.10	10	0.001	0.0001
		Sum	0.5571

$$L_{10} = \frac{10,000}{0.5571}$$

$$= 17,950 \text{ hr}$$

Note that this is the identical prediction obtained by the previous method. As a *user* of traction drives you have at your disposal a method that gives results identical to those the manufacturer of the drive will use in obtaining their basic rating or rating for any combination of different operating conditions and times. The only knowledge that you require beyond the manufacturer's rating and your own intended usage conditions is that the drive *must* have an internal mechanism for adjusting the contact force between its traction elements to be proportional to the load being transmitted. If the design you intend to use does not have that feature, you must calculate on the basis that it is transmitting at full load at all times, and that you may not overload it at all except at your own peril.

Special Case of Drive Known to Have Cylindrical Drive Contacts

In Section 7.3.1 it was pointed out that the life of a traction drive with elliptical contact form was inversely proportional to contact force cubed.

Life

In the general case of Section 7.2.1 it was shown that in elliptical contact cases, the peak compressive stress $S_{c\ max}$ was proportional to the one-third power of contact force. Note that in Section 7.2.1:

$$c = (f) P^{1/3}$$

$$d = (f') P^{1/3}$$

$$\begin{aligned}S_{c\ max} &= \frac{1.5 P}{cd}\\ &= \frac{(f'') P^{3/3}}{P^{1/3} \times P^{1/3}}\\ &= (f'') P^{1/3}\end{aligned}$$

If both sides of this relationship are cubed,

$$S_{c\ max}^{3} = (f''') P$$

If both sides are cubed again,

$$S_{c\ max}^{9} = (f(iv)) P^{3}$$

It follows, in a general way, that if fatigue life of the contact is inversely proportional to contact force P cubed, then it must also be inversely proportional to maximum contact stress to the ninth power. As was shown in Section 7.3.1, the actual stress that is operative in determining the probability of failure is the maximum subsurface shear stress, and the constant of proportionality involves the stressed volume. However, the shear stress and the compressive stress are related, so that the generalization above is valid within a given contact geometry.

Equation (7.2), which is frequently used in conjunction with cylindrical contacts, shows that the maximum contact stress in cylinders rolling on cylinders is proportional to the square of the contact force P. Only one dimension of a cylindrical contact changes as a function of contact force. The dimension transverse to rolling is equal to the cylindrical roller width independent of load.

If this logic were carried to its conclusion, then life of a traction contact with cylindrical rollers, being inversely proportional to $S_{c\ max}$ to the ninth power, would be inversely proportional to the contact force to the 9/2 power. Such a high load/life inverse exponent is never experienced in practice. There are several reasons for this. The most important is that cylindrical contacts are almost always crowned slightly to avoid severe edge stresses and lubrication problems. The crowning creates a certain amount of ellipticity in the contact.

It has been accepted practice in the roller bearing industry to accept the inverse life exponent 10/3 to relate the probably fatigue life of a roller bearing contact to the contact force [8]. The stresses in traction drives with cylindrical contacts are sufficiently similar to those in roller bearings that reading the inverse life exponent across is valid.

Therefore, when the user is examining a traction drive known to have cylindrical contacts, it is appropriate to follow the example of Section 7.3.4 except to use the power 10/3 in the third column of the table.

EXAMPLE. Suppose that a traction drive known to have cylindrical rollers is rated by its manufacturer as having a 10,000-hr life at rated load. It is desired to operate the drive at the same load spectrum as in the General Case of Section 7.3.4.

Construct the following table:

Step description (load fraction)	Time percent at that step	Contact life inverse ratio (step descr. $^{10/3}$)	Contact life inverse ratio × time percent at that step
1.25	10	2.104	0.2104
1.00	20	1.000	0.2000
0.80	30	0.4753	0.1426
0.30	30	0.0181	0.0054
0.10	10	0.0005	0.00005
			0.5584

$$L_{10} = \frac{10,000}{0.5584}$$

$$= 17,908 \text{ hr}$$

The result of this analysis using the 10/3 power is practically the same as the result of the comparable analysis in Section 7.3.4, which was based on the use of the third power for elliptical contacts. The cylindrical contact method shows slightly greater proportion of life used up in the overload period, and conversely, slightly greater life increase in the reduced-load periods. A user who simply adopts the 10/3 power for column 3 of the standard traction drive analysis method will be reasonably close to an accurate evaluation regardless of the exact nature of the drive contacts being analyzed in each case.

REFERENCES

1. R. J. Roark and W. G. Young, *Formulas for Stress and Strain*, 5th ed., McGraw-Hill Book Company, New York.
2. New Departure Engineering Data. Analysis of Stresses and Deflections, New Departure Division, General Motors Corp., 1946, pp. 16, 17.
3. E. N. Bamberger et al., "Life Adjustment Factors for Ball and Roller Bearings—An Engineering Guide," American Society of Mechanical Engineers, New York, 1971.
4. G. Lundberg and A. Palmgren, "Dynamic Capacity of Rolling Bearings," Inginoersventenskapsakademien, Handlingar, no. 196, 1947.
5. D. A. Rohn, S. H. Loewenthal, and J. J. Coy, "Simplified Fatigue Life Analysis for Traction Drive Contacts," American Society of Mechanical Engineers, *J. Mech. Des.*, vol. 103, no. 2, Apr. 1981.
6. J. Lieblein and M. Zelen, "Statistical Investigation of the Fatigue Life of Deep Groove Ball Bearings," Research Paper 2719, *J. Res. National Bureau of Standards*, vol. 57, no. 5, Nov. 1956.
7. E. Shube, "Ball Bearing Survival," *Machine Design*, July 19, 1962. Penton Publishing Company.
8. "Load Ratings and Fatigue Life for Roller Bearings," American National Standard, AFBMA Standard, STD 11-1978, Nov. 21, 1978.

8
CHARACTERISTICS OF POWER SOURCES

8.1 INTRODUCTION

The objective of inclusion of chapters on characteristics of power sources and characteristics of loads in this book, together with the characteristics of various types of drives, is to enable the engineer to select a drive that will couple the power source selected with the load to obtain a result that will be friendly to all three—and thus to the engineer's reputation.

The vast majority of traction drives used in industry are powered by commercial ac motors. These can be treated in considerable depth because of the standardization of available characteristics by NEMA [1] in the United States or by IEC [2] in Europe.

Some traction drives are powered by other sources, such as hydraulic motors, spark or compression ignition engines, turbines, or dc motors. These have not been subject to such comprehensive standardization as have the ac motors. Treatment of them has not been attempted in this book.

8.2 ELECTRIC MOTORS

Electric motors may differ from one another in frame size, frame type, line voltage, number of phases, insulation rating, starting means, torque-speed characteristic, and duty rating. One source [3] has estimated that a 10-hp electric motor can be built in over 40,000 different combinations of electrical and mechanical characteristics.

For proper application to a drive system it is as important to know the characteristics of the motor as it is to know the characteristics of the load and of the drive. As noted earlier, such organizations as

Electric Motors 207

NEMA [1] in the United States and IEC [2] in Europe have helped by standardizing available motor characteristics. The NEMA standards are exclusively expressed in inch derived units. The IEC standards are, of course, metric. Knowing both standards will enable the user to adapt motors made in one system to drives made in the other.

8.2.1 Frame Sizes

NEMA motors are standardized into two- or three-digit frame numbers. Given a frame number, all the critical installation dimensions are established. Two motors with the same frame number will be interchangeable *physically*. Frame number does not relate directly to any performance characteristic.

Two-digit NEMA frame numbers, which are used only for some fractional-horsepower motors, are keyed to the dimension relating the center of the shaft to the bottom of the mounting feet of foot-mounted motors. The two-digit frame size is this dimension expressed in sixteenths of an inch.

Three-digit NEMA frame numbers are similarly keyed, except that only the first two digits of the frame number are part of the key. These express the dimension from the center of the shaft to the bottom of the mounting feet of foot-mounted motors in quarters of an inch. The third digit keys the axial distance between mounting feet to a tabulated value.

Thus a NEMA 56 frame will have dimension D from the shaft centerline to the mounting foot of 3.5 in. A NEMA 324 frame will have dimension D equal to 8 in. Motor diameter is approximately twice dimension D.

IEC standard motors are organized into frame numbers on the same basis, except that the entire frame number represents the distance from the center of the shaft to the bottom of the mounting feet in millimeters. This dimension is universally coded H on IEC diagrams. An IEC 112 frame motor would have dimension H equal to 112 mm (4.409 in.).

Table 8.1 lists the NEMA and IEC frame numbers that are available in each system, and their dimensions.

8.2.2 Standard Motor Horsepowers

Certain standard nominal values of output horsepower are available from NEMA motors. These are keyed to specific frame sizes by NEMA Standard MG1, Part 13. Tables 8.2-8.4 summarize the standard horsepower levels and their associated frames for open-frame single-phase, open-frame polyphase, and totally enclosed polyphase ac motors. There are only small differences in the frame numbers of Tables 8.3 and 8.4.

TABLE 8.1 Standard Motor Frame Sizes[a]

NEMA frame number[b]	Dimension D (in.)	IEC frame number	Dimension H (mm)	(in.)
42	2.63	56	56	(2.20)
48	3.00	63	63	(2.48)
56	3.50	71	71	(2.80)
66	4.13	80	80	(3.15)
14-	3.50	90	90	(3.54)
18-	4.50	100	100	(3.94)
21-	5.25	112	112	(4.41)
25-	6.25	132	132	(5.20)
28-	7.00	160	160	(6.30)
32-	8.00	180	180	(7.09)
36-	9.00	200	200	(7.87)
40-	10.00	225	225	(8.86)
44-	11.00	250	250	(9.84)
50-	12.50	280	280	(11.02)
58-	14.50	315	315	(12.40)
68-	17.00	355	355	(13.98)
		400	400	(15.75)
		450	450	(17.72)

[a]Other, larger, frame sizes exist in both systems.
[b]The dash following the first two digits indicates that several third digits are available in this frame size. The third digit codes the axial distance between mounting holes of foot-mounted machines in accordance with tables that may be found in NEMA Standard MG1, Part 11.

Electric Motors

TABLE 8.2 Frame Designations for Single-Phase Design L Horizontal and Vertical Motors, 60-Hz, Class B Insulation System, Open Type, 1.15 Service Factor, 230 V and Less

Hp	Speed, Rpm		
	3600	1800	1200
¾			145T
1	...	143T	182T
1½	143T	145T	184T
2	145T	182T	...
3	182T	184T	...
5	184T	213T	...
7½	213T	215T	...

Note: See MG 1-11.31 in NEMA Publication No. MG 1 for the dimensions of the frame designations. Suggested Standard for Future Design 7-7-1965, revised 11-11-1965, NEMA Standard 7-16-1969, revised 1-17-1974.
Source: NEMA MG 13-1.01; courtesy of the National Electric Manufacturers Association, New York.

8.2.3 Mounting

There are several ways that the electric motor can be mounted to the drive. Types C, D, J, and N motors are flange mounted. The motor mounting face has a pilot diameter which maintains the motor concentric with the drive. Type C is mounted by tapped holes in the motor face. Type D is mounted by bolts that pass through clearance holes in the flange. Type J is more or less identical with type C except that the shaft end is tapered to accept direct-mounted impellers or similar loads. The shaft end is threaded. Type N is a special mount for oil burner motors. It uses only two mounting holes which pass through ears on the flange.

The other NEMA motors are foot mounted. When foot-mounted motors are used, the achievement of the necessary alignment between the motor and the drive is left to the user. Tolerances of the D (NEMA) and H (IEC) dimensions are such that a shim may be required between the mounting foot and the base to achieve the nominal dimension. Dimensions D and H cannot be larger than the nominal. This does not assist with lateral alignment, of course. The user may want to select a reliable flexible coupling for installation between the motor and the drive. The coupling should have offset capability to account for possible lateral and vertical misalignment.

NEMA motors are not available in skirt-mounted styles. The "skirt," if needed, would be a separate adapter piece between the motor and the drive. Skirt mountings are available as part of the motor from IEC sources.

TABLE 8.3 Frame Designations for Polyphase Squirrel-Cage Designs A and B Horizontal and Vertical Motors, 60 Hz, Class B Insulation System, Open Type, 1.15 Service Factor, 575 V and Less*

Hp	Speed, Rpm			
	3600	1800	1200	900
½	...	...	...	143T
¾	...	...	143T	145T
1	...	143T	145T	182T
1½	143T	145T	182T	184T
2	145T	145T	184T	213T
3	145T	182T	213T	215T
5	182T	184T	215T	254T
7½	184T	213T	254T	256T
10	213T	215T	256T	284T
15	215T	254T	284T	286T
20	254T	256T	286T	324T
25	256T	284T	324T	326T
30	284TS	286T	326T	364T
40	286TS	324T	364T	365T
50	324TS	326T	365T	404T
60	326TS	364T†	404T	405T
75	364TS	365T†	405T	444T
100	365TS	404T†	444T	445T
125	404TS	405T†	445T	...
150	405TS	444T†	...	...
200	444TS	445T†	...	...
250 ‡	445TS	...	...	...

*The voltage rating of 115 V applies only to motors rated 15 hp and smaller.
†When motors are to be used with V-belt or chain drives, the correct frame size is the frame size shown but with the suffix letter S omitted. For the corresponding shaft extension dimensions, see MG 1-11.31 in NEMA Publication No. MG1.
‡The 250 hp rating at the 3600 rpm speed has a 1.0 service factor.
Note: See MG 1-11.31 in NEMA Publication No. MG 1 for the dimensions of the frame designations. Suggested Standard for Future Design 1-21-1964, revised 11-12-1964; 7-7-1965; 11-11-1965; 8-20-1966, NEMA Standard 7-16-1969, revised 1-17-1974.
Source: NEMA MG 13-1.02; courtesy of the National Electrical Manufacturers Association, New York.

TABLE 8.4 Frame Designations for Polyphase Squirrel-Cage Designs A and B Horizontal and Vertical Motors, 60 Hz, Class B Insulation System, Totally Enclosed Fan-Cooled Type, 1.00 Service Factor, 575 V and Less*

Hp	Speed, Rpm			
	3600	1800	1200	900
½	...	...	...	143T
¾	...	...	143T	145T
1	...	143T	145T	182T
1½	143T	145T	182T	184T
2	145T	145T	184T	213T
3	182T	182T	213T	215T
5	184T	184T	215T	254T
7½	213T	213T	254T	256T
10	215T	215T	256T	284T
15	254T	254T	284T	286T
20	256T	256T	286T	324T
25	284TS	284T	324T	326T
30	286TS	286T	326T	364T
40	324TS	324T	364T	365T
50	326TS	326T	365T	404T
60	364TS	364TS†	404T	405T
75	365TS	365TS†	405T	444T
100	405TS	405TS†	444T	445T
125	444TS	444TS†	445T	...
150	445TS	445TS†	...	...
200	...	...	...	...
250	...	...	...	...

*The voltage rating of 115 V applies only to motors rated 15 hp and smaller.
†When motors are to be used with V-belt or chain drives, the correct frame size is the frame size shown but with the suffix letter S omitted. For the corresponding shaft extension dimensions, see MG 1-11.31 in NEMA Publication No. MG 1.
Note: See MG 1-11.31 in NEMA Publication No. MG 1 for the dimensions of the frame designations. Suggested Standard for Future Design 1-21-1964, revised 11-12-1964; 7-7-1965, 11-11-1965, NEMA Standard 7-16-1969, revised 1-17-1974.
Source: NEMA 13-1.03; courtesy of the National Electrical Manufacturers Association, New York.

TABLE 8.5 NEMA C Flange Motors: Mounting Details[a]

Frame size[b]	Pilot diameter AK	Tapped hole BC AJ	Number of holes BD	Tap Size BD	Tap depth BD	Shaft diameter U	Shaft length AH	Key width S	Key length ES
42C	3.00	3.75	4	$\frac{1}{4}$-20	–	0.375	1.312	–	–
48C	3.00	3.75	4	$\frac{1}{4}$-20	–	0.500	1.69	–	–
56C	4.50	5.875	4	$\frac{3}{8}$-16	–	0.625	2.06	0.188	1.41
14-TC	4.50	5.875	4	$\frac{3}{8}$-16	0.56	0.875	2.12	0.188	1.41
18-TC	8.50	7.250	4	$\frac{1}{2}$-13	0.75	1.125	2.62	0.250	1.78
18-TCH	4.50	5.875	4	$\frac{3}{8}$-16	0.56	1.125	2.62	0.250	1.78
21-TC	8.50	7.250	4	$\frac{1}{2}$-13	0.75	1.375	3.12	0.312	2.41
25-TC	8.50	7.250	4	$\frac{1}{2}$-13	0.75	1.625	3.75	0.375	2.91
28-TC	10.50	9.00	4	$\frac{1}{2}$-13	0.75	1.875	4.38	0.500	3.28
28-TSC	10.50	9.00	4	$\frac{1}{2}$-13	0.75	1.625	3.00	0.375	1.91
32-TC	12.50	11.00	4	$\frac{5}{8}$-11	0.94	2.125	5.00	0.500	3.91
32-TSC	12.50	11.00	4	$\frac{5}{8}$-11	0.94	1.875	3.50	0.500	2.03
36-TC	12.50	11.00	8	$\frac{5}{8}$-11	0.94	2.375	5.62	0.625	4.28
36-TSC	12.50	11.00	8	$\frac{5}{8}$-11	0.94	1.875	3.50	0.500	2.03
40-TC	12.50	11.00	8	$\frac{5}{8}$-11	0.94	2.875	7.00	0.750	5.65
40-TSC	12.50	11.00	8	$\frac{5}{8}$-11	0.94	2.125	4.00	0.500	2.78
44-TC	16.00	14.00	8	$\frac{5}{8}$-11	0.94	3.375	8.25	0.875	6.91
44-TSC	16.00	14.00	8	$\frac{5}{8}$-11	0.94	2.375	4.50	0.625	3.03
50-	16.50	14.50	4	$\frac{5}{8}$-11	0.94	–	–	–	–

[a] All dimensions in inches.
[b] The dash in the three-digit frame sizes can stand for any third digit. The third digit keys axial length dimensions, which are not addressed in this table. For example, the dimensions given for frame size 14-TC apply to frame sizes 143TC, 145TC, etc.

TABLE 8.6 Dimensions of IEC Style D Flange-Mounted Metric Frames[a]

Frame size	(kW) Output rating[b] 3000 rpm	1500 rpm	1000 rpm	Pilot diameter N	Bolt hole BC	Number of holes	Hole size	Shaft diameter	Shaft length	Keyway Width	Length
90	1.5	1.1	0.75	130	165	4	12	24	50	8	40
90	2.2	1.5	1.1	130	165	4	12	24	50	8	40
100	3.	2.2	1.5	180	215	4	15	28	60	8	45
112	4.	4.	2.2	180	215	4	15	23	60	8	45
132	7.5	5.5	3.	230	265	4	15	38	80	10	60
160	18.	15.	11.	250	300	4	19	42	110	12	85
180	22.	18.5	15.	250	300	4	19	48	110	14	90
200	37.	30.	22.	300	350	4	19	55	140	16	90
225	45.	—	—	350	400	8	19	55	140	16	90
225	—	45.	30.	350	400	8	19	60	110	18	120
250	55.	—	—	450	500	8	19	60	140	18	120
250	—	55.	37.	450	500	8	19	65	140	18	120
250	—	75.	45.	450	500	8	19	70	140	20	120
280	75	—	—	450	500	8	19	65	140	18	120
280	—	75.	45.	450	500	8	19	75	140	18	120
280	—	90.	55.	450	500	8	19	80	170	22	120

[a]All dimensions in millimeters. Multiply millimeters by 0.03937 to obtain inch equivalent; multiply kilowatts by 1.34 to obtain horsepower equivalent.
[b]Some frame sizes offer several shafts, depending on kilowatt output rating. The kilowatt output rating columns have no particular significance for frame sizes 90 to 200.

Tables 8.5 to 8.8 list the dimensions of the NEMA C flange mountings, the IEC D flange mountings, the IEC C flange mountings, and the IEC skirt-mounted motors, respectively.

8.2.4 Enclosures

Motors have important construction features which must be specified aside from how the motor is to be mounted. For example, motors can be open, drip-proof, totally enclosed fan-cooled, or explosion-proof, depending on the degree of environmental protection required. Explosion-proof motors look very much the same as totally enclosed fan-cooled motors except that the length of the lap joints between the housing parts has been made great enough to quench any explosion (deliberately created within the motor during the laboratory certification process) before it could ignite a flammable mixture surrounding the motor. The open motor is the lowest-cost construction. It should not be used where its cooling passages could get clogged with lint or where water or other harsh environment elements get access to the electrical parts. Figures 8.1 to 8.5 illustrate some of these motor mechanical features.

TABLE 8.7 Dimensions of IEC Style C Face-Mounted Metric Frames[a]

Frame size	Pilot diameter N	Pitch circle diameter	Number of tapped holes on PC	Thread size	Shaft diameter	Shaft length	Keyway Width	Keyway Length
56	50	65	4	M5	–	–	–	–
63	60	75	4	M5	14	30	5	20
71	70	85	4	M6	14	30	5	20
80	80	100	4	M6	19	40	6	25
90	95	115	4	M8	24	50	8	40
100	110	130	4	M8	28	60	8	45
112	110	130	4	M8	28	60	8	45
132	130	165	4	M10	38	80	10	60
160	180	215	4	M12	42	110	12	85

[a]All dimensions in millimeters. Multiply millimeters by 0.03937 to obtain inch equivalent.

Electric Motors

TABLE 8.8 Dimensions of IEC Skirt Frames[a]

Frame size	Pilot diameter, N	Pilot length, T	Bolt circle diameter	Number of bolt holes	Hole size	Skirt offset, R	Diameter	Keyway Length	Keyway Width	Keyway Length
80	130	3.5	165	4	12	53	19	40	6	25
90	130	3.5	165	4	12	53	24	50	8	40
100	180	4.0	215	4	15	63	28	60	8	45
112	180	4.0	215	4	15	63	28	60	8	45
132	230	4.0	265	4	15	83	38	80	10	60
160	250	5.0	300	4	19	113	42	110	12	85
180	250	5.0	300	4	19	113	48	110	14	90
200	300	5.0	350	4	19	113	55	140	16	90
225	350	5.0	400	8	19	143	60	140	18	110
250	450	5.0	500	8	19	173	65	140	18	110
280	450	5.0	500	8	19	173	75	140	20	110
315	550	6.0	600	8	24	173	80	170	22	140
355	680	6.0	740	8	24	213	–	–	–	–
400	880	6.0	940	8	28	213	–	–	–	–

[a]All dimensions in millimeters. Multiply millimeters by 0.03937 to obtain inch equivalent.

For size 225 and larger motors that run at 1500 rpm or less, use the size shown. For motors that run at 3000 rpm, use the shaft listed for the next smaller frame.

This table is valid for totally enclosed fan-cooled motors, and air-over motors of all sizes. It may be used for open motors smaller than frame size 160. For open motors of frame size 160 or larger, select the shaft size corresponding to one frame size larger than would have been selected in accordance with the note above.

Figure 8.1 shows a foot-mounted open motor, whereas Fig. 8.2 illustrates a foot-mounted totally enclosed fan-cooled motor. The fan draws air in through the rear louvers, around the body of the motor—which has no openings—and out the front. (NEMA standards refer to the shaft end of the motor as the rear and the closed end as the front. It does not matter which end is called what. The only cause for caution is when direction of rotation is specified in terms of "front" or "rear.") The shell shown in this picture is one that surrounds the motor and not the motor body itself.

Figure 8.3 is a three-phase C flange-mounted totally enclosed fan-cooled motor. The cooling is the same as the motor in Fig. 8.2 but the mounting is quite different.

FIG. 8.1 Foot-mounted open motor. (Courtesy of Westinghouse Small Motor Division, Lima, Ohio.)

Electric Motors 217

FIG. 8.2 Foot-mounted totally enclosed fan-cooled motor. (Courtesy of Westinghouse Small Motor Division, Lima, Ohio.)

Figure 8.4 is a split-phase C flange-mounted explosion-proof motor. Note the rugged construction.

Finally, Fig. 8.5 is a three-phase foot-mounted totally enclosed fan-cooled motor. Note that it has a lifting eye and drain plugs. An explosion-proof motor in this frame size would look no different externally except that the conduit box might appear more rugged. This motor is cooled by air blown over the body by the fan within the shroud. There is no shroud over the motor body.

8.2.5 Torque-Speed Characteristics

Polyphase Induction Motors

In the notation of most engineering graphics the independent variable is displayed along the abscissa and the dependent variable is displayed along the ordinate. In the familiar x,y plot the value of y is a function of x. The conventions of electric motor engineering seem to be independent of such constraints. The most common display format is torque along the horizontal axis, and speed, current, power factor, and efficiency along the vertical axis.

Figure 8.6 is a representation of the torque-speed characteristic of a typical polyphase induction motor. In this case the parameters arise from the NEMA design B characteristic for a 20- to 30-hp four-pole motor which would have a synchronous speed of 1800 rpm at 60 Hz or 1500 rpm at 50 Hz. This figure is normalized in terms of torque expressed as a percentage of full-load torque, and speed expressed as a percentage of synchronous speed.

FIG. 8.3 Three-phase NEMA C flange-mounted totally enclosed fan-cooled motor. (Courtesy of Westinghouse Small Motor Division, Lima, Ohio.)

FIG. 8.4 Split-phase NEMA C flange-mounted explosion-proof motor. (Courtesy of Westinghouse Small Motor Division, Lima, Ohio.)

Consider this motor standing still with no power applied. Suddenly, full-rated voltage is impressed on the motor terminals. The torque developed, which gets the motor started and accelerating, is called the locked-rotor torque or the starting torque. Note that in Fig. 8.6 this value is 150% of the full-load torque rating of this particular motor. The design value of this normalized torque is not independent of motor size or number of poles. Table 8.9 summarizes the values of the locked-rotor torque expressed as a percent of full-load torque and as a function of motor horsepower rating and synchronous speed. As will be discussed further in this section, there are several NEMA design curves: A, B, C, D, and F. The locked-rotor torque for designs A and B are the same and therefore may be included in the same section of the table.

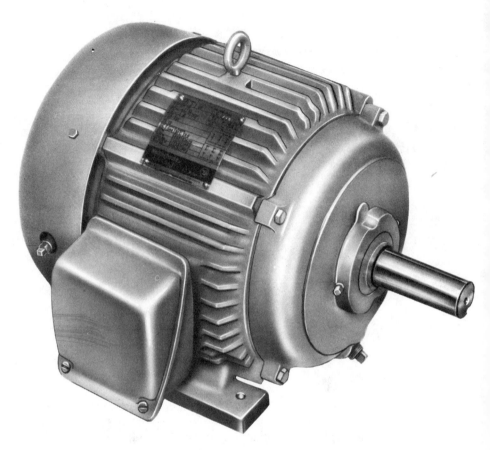

FIG. 8.5 Three-phase foot-mounted totally enclosed fan-cooled motor. (Courtesy of Westinghouse Electric Corp., Buffalo, N.Y.)

When the motor begins to turn, the available torque begins to fall. In the NEMA designs A and B an intermediate minimum torque is reached and the available torque begins to increase as speed is increased. This minimum is called the minimum torque or the pull-up torque. Table 8.10 summarizes the relationship of pull-up torque to either locked-rotor torque or full-load torque, as appropriate.

As the motor speed increases still further, the torque for NEMA designs A and B reaches a maximum. This part of the curve has the shape of a knee. The value of the maximum is called the breakdown torque or the pull-out torque. Table 8.11 summarizes the relationship of breakdown torque to full-load torque.

Electric Motors

As speed is increased further, the torque falls off rapidly. At about 96% of synchronous speed the torque is equal to the full-load torque. This is the point at which motor torque and load torque are in balance and the motor does not accelerate further. The difference between the speed at which full-load torque is developed and synchronous speed is called the "slip." Slip is necessary for the induction motor to develop the rotor currents that interact with the stator flux to produce net torque. If an induction motor could somehow be brought all the way up to synchronous speed it would have no torque at all.

A point could be found on Fig. 8.6 where an imaginary vertical load torque line could be drawn which would intercept the torque-speed characteristic at three speeds. The lowest and highest of the three points are stable. That is, the motor would not be able to accelerate through the lowest intersection because, just above that speed, the motor torque would be less than the load torque. This is also true of the highest intersection.

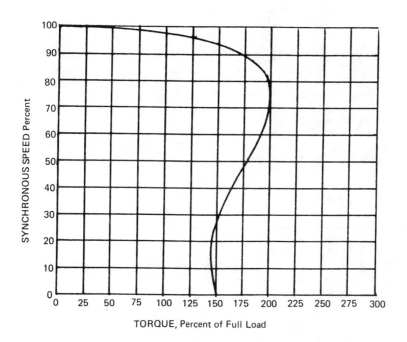

FIG. 8.6 Typical torque-speed characteristics of NEMA design B induction motor. See Tables 8.9-8.11 for critical points. (Courtesy of Westinghouse Small Motor Division, Lima, Ohio.)

TABLE 8.9 Locked-Rotor Torque of Single-Speed Polyphase Squirrel-Cage Integral-Horsepower Motors with Continuous Ratings

A. The locked-rotor torque of Design A and B, 60- and 50-hertz, single-speed, polyphase squirrel-cage motors, with rated voltage and frequency applied, shall be not less than the following values which are expressed in percent of full-load torque. For applications involving higher torque requirements, see the locked-rotor torque values for Design C and D motors.

Hp	Synchronous Speed, Rpm						
	60 hertz 3600	1800	1200	900	720	600	514
	50 hertz 3000	1500	1000	750	...	...	...
½	...	...	...	140	140	115	110
¾	...	...	175	135	135	115	110
1	...	275	170	135	135	115	110
1½	175	250	165	130	130	115	110
2	170	235	160	130	125	115	110
3	160	215	155	130	125	115	110
5	150	185	150	130	125	115	110
7½	140	175	150	125	120	115	110
10	135	165	150	125	120	115	110
15	130	160	140	125	120	115	110
20	130	150	135	125	120	115	110
25	130	150	135	125	120	115	110
30	130	150	135	125	120	115	110
40	125	140	135	125	120	115	110
50	120	140	135	125	120	115	110
60	120	140	135	125	120	115	110
75	105	140	135	125	120	115	110
100	105	125	125	125	120	115	110
125	100	110	125	120	115	115	110
150	100	110	120	120	115	115	...
200	100	100	120	120	115	...	...
250	70	80	100	100	...	...	...
300	70	80	100	...	...	...	...
350	70	80	100	...	...	...	...
400	70	80	...	...	...	...	...
450	70	80	...	...	...	...	...
500	70	80	...	...	...	...	...

B. The locked-rotor torque of Design C, 60- and 50-hertz, single-speed, polyphase squirrel-cage motors, with rated voltage and frequency applied, shall be not less than the following values which are expressed in percent of full-load torque.

Hp	Synchronous Speed, Rpm		
	60 hertz 1800	1200	900
	50 hertz 1500	1000	750
3	...	250	225
5	250	250	225
7.5	250	225	200
10	250	225	200
15	225	200	200
20–200, inclusive	200	200	200

Electric Motors

C. The locked-rotor torque of Design D, 60- and 50-hertz, 4-, 6- and 8-pole, single-speed, polyphase squirrel-cage motors rated 150 horsepower and smaller, with rated voltage and frequency applied, shall be not less than 275 percent, expressed in percent of full-load torque.

NEMA Standard 8-7-1947, revised 6-24-1949; 11-17-1955; 11-17-1966; 7-16-1969; 9-20-1978

Source: NEMA MG 1-12.37; courtesy of the National Electrical Manufacturers Association, New York.

TABLE 8.10 Pull-up Torque of Single-Speed Polyphase Squirrel-Cage Integral-Horsepower Motors with Continuous Ratings, Designs A, B, and C

The pull-up torque of Design A and B, single-speed, polyphase squirrel-cage motors, with rated voltage and frequency applied, shall be not less than the following:

Column 1 Locked-rotor Torque from MG 1-12.37	Column 2 Pull-up Torque, Percent
110 percent or less	90 percent of Column 1
greater than 110 percent but less than 145 percent	100 percent of full-load torque
145 percent or more	70 percent of Column 1

The pull-up torque of Design C motors, with rated voltage and frequency applied, shall be not less than 70 percent of the locked-rotor torque from par. B of MG 1-12.37.

NEMA Standard 8-7-1947, revised 11-13-1958; 11-17-1966; 1-25-1972.

Source: NEMA MG 1-12.39; courtesy of the National Electrical Manufacturers Association, New York.

TABLE 8.11 Breakdown Torque of Single-Speed Polyphase Squirrel-Cage Integral-Horsepower Motors with Continuous Ratings (Revised)

A. The breakdown torque of Design A and B, 60- and 50-hertz, single-speed, polyphase squirrel-cage motors, with rated voltage and frequency applied, shall be not less than the following values which are expressed in percent of full-load torque.

Hp	60 hertz	50 hertz	Synchronous Speed, Rpm					
	60 hertz	3600	1800	1200	900	720	600	514
	50 hertz	3000	1500	1000	750	...	...	...
½	...	...	...	225	200	200	200	
¾	...	...	275	220	200	200	200	
1	...	300	265	215	200	200	200	
1½	250	280	250	210	200	200	200	
2	240	270	240	210	200	200	200	
3	230	250	230	205	200	200	200	
5	215	225	215	205	200	200	200	
7½	200	215	205	200	200	200	200	
10–125, inclusive	200	200	200	200	200	200	200	
150	200	200	200	200	200	200	...	
200	200	200	200	200	200	...	...	
250	175	175	175	175	...	...	...	
300–350	175	175	175	...	...	...	...	
400–500, inclusive	175	175	...	...	...	...	...	

B. The breakdown torque of Design C, 60- and 50-hertz, single-speed, polyphase squirrel-cage motors, with rated voltage and frequency applied, shall be not less than the following values which are expressed in percent of full-load torque.

Hp	Synchronous Speed, Rpm			
	60 hertz	1800	1200	900
	50 hertz	1500	1000	750
3		...	225	200
5		200	200	200
7½–200, inclusive		190	190	190

NEMA Standard 1-26-1948, revised 6-24-1949; 11-17-1955; 11-17-1966; 7-16-1969; 9-20-1978.

Source: NEMA MG 1-12.38; courtesy of the National Electrical Manufacturers Association, New York.

Electric Motors

The low-speed end of the curve represents an area of severe heating of the rotor by the high rotor bar currents induced by the high slip frequency. The motor has to go through this part of the curve on every startup, but should never be allowed to dwell there. It is important that the load placed on the motor during startup always be less than the pull-up torque rating so that the motor is free to accelerate rapidly through this high-slip low-efficiency condition.

NEMA has made provision for five recognized induction motor torque-speed characteristics, designs A, B, C, D, and F. The critical points have been tabulated in Tables 8.9-8.11. These curves are plotted on Figs. 8.7, 8.8, 8.9, and 8.10 for NEMA designs A, C, D, and F, respectively. These differences are achieved by changes in rotor bar resistance and shape, in conjunction with stator winding changes. Table 8.12 summarizes differences between the characteristics and when each is likely to be applied. Design D is of particular utility with severe intermittent loads such as those encountered with punch presses, in which energy is drawn from a flywheel, which slows down

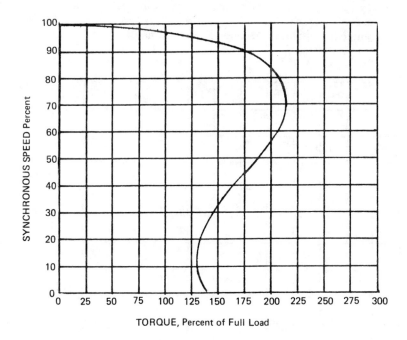

FIG. 8.7 Typical torque-speed characteristics of NEMA design A induction motor. See Tables 8.9-8.11 for critical points. (Courtesy of Westinghouse Small Motor Division, Lima, Ohio.)

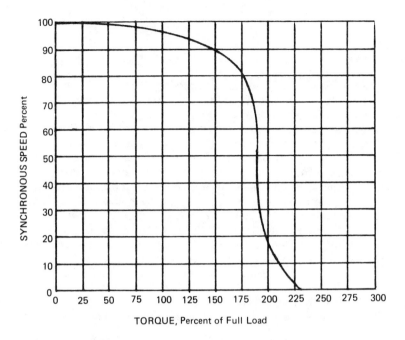

FIG. 8.8 Typical torque-speed characteristics of NEMA design C induction motor. See Tables 8.9-8.11 for critical points. (Courtesy of Westinghouse Small Motor Division, Lima, Ohio.)

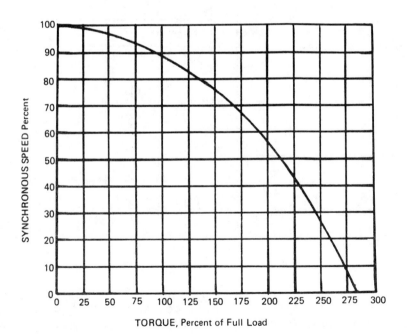

FIG. 8.9 Typical torque-speed characteristics of NEMA design D induction motor. See Table 8.9C for locked-rotor torque. (Courtesy of Westinghouse Small Motor Division, Lima, Ohio.)

Electric Motors

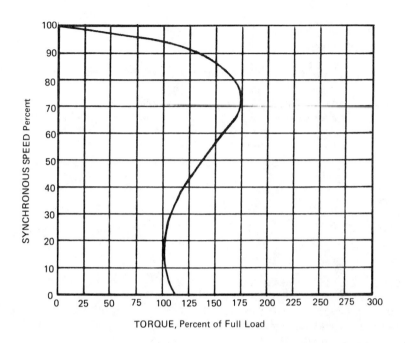

FIG. 8.10 Typical torque-speed characteristics of NEMA design F induction motor. (Courtesy of Westinghouse Small Motor Division, Lima, Ohio.)

TABLE 8.12 Overview of NEMA Design Characteristics and Applications

Design	Pull-up torque	Starting current	Breakdown torque	Full-load slip	Typical application
A	Normal	Normal	High	Low	Machine tools, centrifugal pumps, fans
B	Normal	Low	High	Low	Same as design A
C	High	Low	Normal	Low	Compressors, conveyers
D	Very high	Low	—	High	Punch presses
F	Low	Very low	Very low	High	Fans (large)

Source: Courtesy of the Westinghouse Electric Corporation, Buffalo, N.Y.

and must be brought back up to speed by the motor, which should not be severely overloaded by the speed change.

Rotor Inertias. Chapter 12 contains several illustrations as to how the motor torque is distributed during starting transients. The value of motor torque during a startup may prove to be a more important factor in the selection of a traction drive than the value of the steady-state torque. During startup the motor must accelerate its own rotor as well as the connected load. Because of this the load never sees the full starting torque of the motor. The torque available to accelerate the load is

$$T_{\text{acceleration load}} = (T_{\text{motor}} - T_{\text{load friction}}) \frac{Wk^2_{\text{load}}}{Wk^2_{\text{motor}} + Wk^2_{\text{load}}}$$

(8.1)

The evaluation process described in Chapter 12 can proceed only if the motor Wk^2 is known. When a final motor selection is made the manufacturer should be consulted so that the actual rotor inertia may be stated. For preliminary design purposes Fig. 8.11 will prove useful. Figure 8.11 presents typical rotor inertias of motors from 1 to 300 hp as a function of the number of poles. [Motor nominal speed is 60 × frequency (Hz)/half the number of poles.]

Load Inertias. The maximum inertia that can be started by a polyphase motor is limited by its internal heating which takes place in the starting process. NEMA, in MG 1-20.42, has published a table of maximum load inertias as a function of motor horsepower and speed. The table was calculated from the formula

$$\text{Load } Wk^2 = A \left| \frac{hp^{0.95}}{(rpm/1000)^{2.4}} \right| - 0.0685 \left| \frac{hp^{1.5}}{(rpm/1000)^{1.8}} \right| \quad (8.2)$$

where A is 24 for 300 to 1800-rpm motors and 27 for 3600-rpm motors.

Figure 8.12 will prove helpful in estimating whether the inertia of an intended load exceeds the allowable rating of a candidate motor. If the values seem close, or if the number of starts is frequent, the motor manufacturer should be consulted for a recommendation.

Single-Phase AC Motors

Fractional-horsepower and small integral-horsepower ac motors are sometimes supplied as single-phase units. These require only two supply wires and are easily switched on and off by simple controls.

Electric Motors

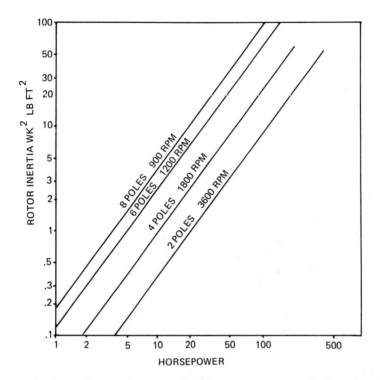

FIG. 8.11 Typical inertia values of rotors of frames 182 to 445 NEMA series three-phase induction motors continuous ratings on 60-Hz power. (Courtesy of Westinghouse Electric Corp., Buffalo, N.Y.)

A single-phase induction motor will run at low slip at rated load just like its polyphase cousin. The problem is getting it started, as it has zero torque at zero speed. If you had an unloaded single-phase ac motor without any starting provision (e.g., one with a disconnected starting winding) you could connect it to a power line and start it in either direction of rotation with a flip of the wrist and it would come up to speed in the direction in which it was initially started.

The technique used to start single-phase motors is to introduce an auxiliary magnetic field which either leads or lags the main magnetic field by a sufficient electrical angle so that the field rotation sequence is unambiguous. This is done by installing a starting winding and adjusting the reactance of the starting winding circuit so that the phase of the current in the winding is displaced about the same electrical angle with respect to the current in the main winding as the

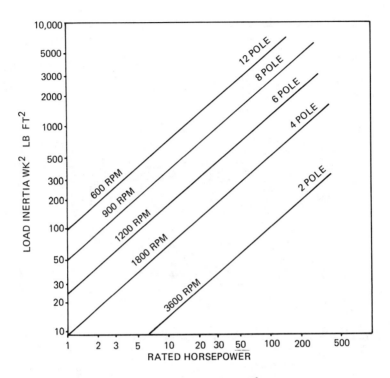

FIG. 8.12 Allowable connected load Wk^2 for squirrel-cage induction motors, three-phase ac 60-Hz frames 143T to 449T. (Courtesy of Westinghouse Electric Corp., Buffalo, N.Y.)

starting winding coils are shifted mechanically in the stator slots with respect to the main winding. This is usually done by one of three methods: resistance split phase, capacitance split phase, or reluctance techniques.

Resistance split phase. This type of motor has a starting winding, usually of much finer wire than the main winding, connected to the motor terminals by a centrifugal switch. The phase angle of the current in this winding, which depends on the ratio of its resistance to its reactance, is different from that of the current in the low-resistance main winding. When the motor reaches about three-fourths of synchronous speed the centrifugal switch opens and the motor continues to accelerate to full-load speed on the main winding alone.

The centrifugal switch resets when the motor is next stopped. Reversal of the direction of starting is obtained by reversing the connections of the starting winding with respect to the terminals of the main winding. For this purpose the starting winding leads are brought to a terminal board at the front of the motor. Instructions are pasted to the inside of the terminal box cover or are engraved on the nameplate.

Figures 8.13 and 8.14 show the torque-speed characteristic of two different resistance split-phase motors. The parameters are made different by varying the parameters of the starting winding and the resistance of the rotor bars.

Capacitance split phase. The necessary phase shift of the current in the starting winding can be accomplished by adding an electrolytic or an oil-filled capacitor in series with the winding. This permits a substantial phase shift in a heavy-gauge winding since resistance of the wire is not an important factor. The starting winding is cut out by a centrifugal switch in the same manner as that described above for the resistance split-phase motor. Figure 8.15 shows the torque-speed curve for a typical capacitance split-phase motor.

Permanent capacitor split phase. Some light-duty motors may omit the centrifugal switch and thus leave the capacitor in series with the auxiliary winding across the line at all times. When this is the case, the capacitor is usually smaller than the capacitor used in the capacitor split-phase motor. The torque-speed characteristic curve of such a motor is shown in Fig. 8.16.

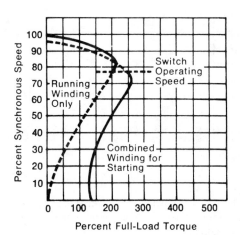

FIG. 8.13 Typical torque-speed characteristics of general-purpose fractional-horsepower resistance split-phase motor. (Courtesy of Westinghouse Small Motor Division, Lima, Ohio.)

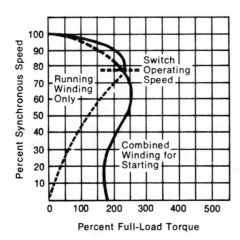

FIG. 8.14 Typical torque-speed characteristics of high-starting-torque fractional-horsepower resistance split-phase motor. (Courtesy of Westinghouse Small Motor Division, Lima, Ohio.)

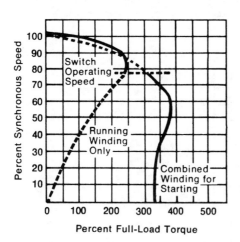

FIG. 8.15 Typical torque-speed characteristics of general-purpose capacitor-start fractional-horsepower motor. (Courtesy of Westinghouse Small Motor Division, Lima, Ohio.)

Electric Motors

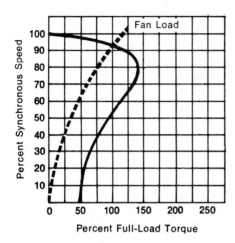

FIG. 8.16 Typical torque-speed characteristics of permanent split-capacitor fractional-horsepower motor. (Courtesy of Westinghouse Small Motor Division, Lima, Ohio.)

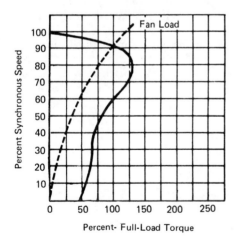

FIG. 8.17 Typical torque-speed characteristics of shaded-pole fractional-horsepower motor. (Courtesy of Westinghouse Small Motor Division, Lima, Ohio.)

Shaded pole. The necessary magnetic field phase lag can be obtained by burying a heavy copper short-circuited turn in one edge of each pole face. The phase relationship between the current in this coil, induced by the alternating flux, and the current in the stator winding is sufficient to cause a preferential rotation to the magnetic field, thereby obtaining unidirectional starting torque. Figure 8.17 illustrates typical performance of this type.

It is not likely that shaded-pole motors will be encountered in traction drive applications. They are the domain of clocks, phonographs, advertising display, fans, and other low-power applications. They cannot be reversed, as the direction of rotation is built into the design.

REFERENCES

1. "Motors and Generators," Publication MG-1, June 1978. National Electrical Manufacturers Association, 2101 L Street, N.W., Washington, DC, 20037.
2. "Dimensions and Output Ratings for Electrical Machines - Frame Numbers 56-400 and Flange Numbers F 55 to F 1080," Publication 72; "Dimensions and Output Ratings for Foot Mounted Electrical Machines with Frame Numbers 355 to 1000," Publication 72A. International Electrotechnical Commission, 1 Rue De Varembe, 1211 Geneva 20, Switzerland.
3. "Motor and Gearing Products, Selection and Application," Publications SA-9916-1 to SA-9916-4. June 1979. Westinghouse Electric Corporation Motor Divisions, P. O. Box 225, Buffalo, NY, 14240.

9
CHARACTERISTICS OF LOADS

9.1 INTRODUCTION

The nature of the load has to be appreciated before the drive which couples it to its power source can be selected properly. There are many more forms and varieties of loads than there are power sources or types of traction drives. The drive specification sheet (Chapter 10) will list certain characteristics of the load which will have to be addressed.

This chapter describes loads in general terms. The applications engineer will know much more about the specific load being driven than can be listed in this book. The information presented in this chapter may be helpful in filling in missing information if the equipment being driven is similar to others of the same generic type and characteristics.

9.2 TORQUE-SPEED

The traction drive may be called on to change speed in response to various conditions of operation. The capability of the drive to provide output torque at different speeds needs to be known. The torque requirement of the driven equipment as a function of speed needs to be compared to the characteristics of the drive to make sure that there are no areas of operation in which the drive is not capable of supporting the load.

Some loads, such as fans, have a rising torque-speed characteristic. In most fans, the torque required is a function of the square of the speed. Because power is the product of torque times speed, fan power is generally a function of the cube of the speed at which the fan is driven. Therefore, the motor power rating and the drive torque rating must be derived for the maximum operating speed of the system. Torque characteristics for centrifugal pumps also vary as the square of the speed. Power varies with the cube of the speed.

Some loads are "constant-horsepower" loads. That is, the torque they require is inverse to their operating speed so that the product of torque times speed is a constant. Some loads with active controls exhibit a constant-horsepower type of characteristic within the limits of the active control capability. For example, a hydraulic pump in which the swash plate angle is automatically varied by an internal mechanism to maintain constant output pressure will automatically destroke if the pump is speeded up, so that the torque on the shaft will be reduced if the speed is increased. This may also occur in generators that have automatic controls which raise or lower the field strength to maintain the voltage constant if the drive is slowed down or speeded up, respectively, within the range of the control.

Other loads may have a flat, or constant-torque, characteristic. That is, the torque may not be particularly related to the speed at which the load is being driven. This may be true of conveyors or hoists or metering mechanisms.

No load will fall exactly into one of these neatly defined characteristics. Generally speaking, mechanical equipment will require more power to drive faster and less power to drive slower just due to internal losses and those associated with moving the load. This chapter makes the point that the complete operating spectrum need be examined to assure that the correct drive-to-load matching point is selected.

9.3 LOAD DIRECTION

Some loads overhaul. When they tend to go faster than the drive which is operating them or the power source that is providing the energy, some real problems could develop if this exigency had not been addressed in the system design. It is easy to point out circumstances in which a load tends to overhaul. Any load with a high inertia, such as a flywheel, will attempt to overhaul the driver if output speed is reduced rapidly. Hoists overhaul when loads are being lowered. Drives used in position devices such as actuators or robots which have to move in one direction against an opposing load sometimes have to move in the other direction and experience an aiding load.

Some traction drives are made so that they cannot support an aiding load. It is a feature of some that there is a single wedge roller which is drawn into contact by the force directions associated with normal load direction. In the reverse load direction the roller is pushed out of contact. This feature may be very important if it is essential that the driven machine never be allowed to reverse even if the phase rotation of the drive is accidentally reversed. This type of drive has an important niche in the list of available types, but it should never be used if the load has any reversing potential at all.

The drive power source may also have to be able to retard any aiding load which is reflected back to it by the drive. Some, such as permanent magnet dc motors and shunt motors, change over to generators if they are pushed faster than they want to go, and apply braking torque to the load. It would be important to check that there are no blocking diodes in the power supply that would prevent this energy from being delivered to the line. Series dc motors and ac induction motors are poor power absorbers. The series motor field strength goes to zero as the current reverses and there is no field left to generate reverse current.

Ac motors are poor absorbers of power. They have to speed up above synchronous speed to become series generators and they do not become good ones at that.

It is important that the application engineer recognize whether the drive system will be overhauled by the load in any operating condition, and if so, where the power is going to go.

9.4 DRIVE DIRECTION

Many loads have to be reversed by the drive system. This can be done by reversing the motor if an electric motor is used, or by reverse gearing if an internal combustion engine is used, or perhaps within the drive itself if its geometry permits. Section 9.3 pointed out one type of drive that could not be reversed intentionally. This paragraph mentions direction reversal only as one characteristic of some load applications. The drive and the power source have to be checked to make sure that compliance with such a requirement creates no special problems.

9.5 CYCLICAL TORQUE VARIATIONS

Very often the load will vary cyclically, yet on average be some steady-state value. Consider a crusher. If a particularly big lump passes through, the drive and the power source will know about it, yet on average the load may be very much less than the occasional peak. Drives have to be sized so that they are not damaged by the peak cyclical load. The power transmission industry uses "service factors" or "shock factors" which in the absence of more specific information

TABLE 9.1 Characteristics of Loads: Service Factors for Torque Peaks

Load type	Service factor	Load type	Service factor
Agitators		Drums, rotating	
Liquid type	1.0	Plating	1.2
Liquid with solids	1.2	Polishing	1.2
Brewery equipment		Tanning	1.2
Bottling machinery	1.2	Elevators	
Cookers	1.0	Belt buckets	1.0
Crushers	1.0	Chain buckets	1.5
Hoppers	1.0	Hoists	1.0
Vats	1.0	Lifts	1.0
Brick and clay machinery		Food machinery	
Clay mixers	1.5	Canning machinery	1.0
Molding machines	1.5	Cookers	1.2
Presses	1.5	Kneading troughs	1.0
Cement machinery		Meat cutters	1.0
Crushers	2.0	Meat grinders	1.2
Dryers	1.5	Root cutters	1.5
Kilns	1.5	Hoists	
Screens		Lifting gears	1.0
Rotary type	1.0	Pile driving	1.5
Shaking type	1.5	Pulling	1.2
Conveyers		Skip	1.5
Apron	1.0	Sluice gates	1.2
Belt	1.0	Lifting machinery	
Bucket	1.5	Lifting gear	1.0
Chain	1.2	Moving horizontally	1.0
Platform	1.5	Slewing	1.0
Roller	1.5	Machine tools	
Screw	1.5	Boring machines	1.0
Vibration	1.8	Drill presses	1.5
Crushers		Lathes	1.0
Ball type	1.5	Milling machines	1.0
Grindstone type	1.0	Planers	1.5
Pebble type	1.5	Punch presses	3.0
Roller type	1.5		
Dredges, bucket chain type	2.0		

Cyclical Torque Variations

Load type	Service factor	Load type	Service factor
Metalworking machinery		Rubber making machinery	
Benders	1.5	Calenders	1.2
Coilers	1.0	Cutting machines	1.2
Drawing bench (wire)	1.0	Extruders	1.0
Ladles	1.2	Mixers	1.5
Mixers	1.5	Roller mills	1.2
Rivet- or chain-making machinery	3.0	Saws	
Rolling mills	2.0	Band saws	1.0
Slitters	1.2	Circular saws	1.0
Spoolers	1.0	Reciprocating saws	1.5
Papermaking machinery		Textile machinery	
Bleachers	1.0	Batchers	1.2
Calenders	1.2	Calenders	1.2
Coilers	1.0	Cards	1.2
Conical refiners	1.2	Coating machines	1.0
Conveyers	1.0	Drums	1.2
Defibering machinery	1.2	Drying cans	1.2
Felt stretchers	1.2	Dyeing machinery	1.2
Guillotines	2.0	Fulling mills	1.5
Mixers	1.2	Impregnation machines	1.0
Paste presses	1.2	Mangles	1.2
Pressing rolls	1.2	Nappers	1.2
Refining piles	1.0	Power looms	2.5
Shredding machines	1.5	Sanforizers	1.5
Spoolers	1.2	Spinners	1.2
Winders	1.0	Spoolers	1.0
Petroleum industry		Tenter frames	1.2
Drills	2.0	Washing cylinders	1.5
Paraffin presses	1.5	Tumbling barrels	
Slush pumps	2.5	Plating	1.2
Printing presses	1.0	Polishing	1.2
		Tanning	1.2
Pumps		Washing machines	
Centrifugal	1.0	Continuous drive	1.0
Gear	1.0	Reversing drive	1.5
Multicylinder reciprocating	2.0		
Single-cylinder reciprocating	3.0		

Source: Courtesy of Plessey Dynamics Corp., Hillside, N.J.

TABLE 9.2 Service Classifications for Various Machines

Type of service	Number of starts per day				
	1-10	11-50	51-200	201-5000	Over 5000
Light	1.0	1.1	1.2	1.3	1.4
Uniform	1.2	1.3	1.4	1.6	1.7
Moderate	1.5	1.6	1.8	2.0	2.1
Severe	2.0	2.2	2.4	2.6	2.8
Very severe	2.5	2.7	3.0	3.3	3.5

Source: Courtesy of Plessey Dynamics Corp., Hillside, N.J.

enable a certain margin of safety to be built into the design selection based on the type of load and the number of starts and stops. Clearly, centrifugal pumps would require much less design margin for cyclical torque peaks than crushers or paper-cutting guillotines.

Table 9.1 is a list of many general applications with a service factor tabulated for each. Occasionally, a similar list will be encountered in terms of "uniform load," "light shock," "moderate shock," "severe shock," and "very severe shock." Table 9.2 lists a rough correspondence between these descriptions and the service factors tabulated in Table 9.1.

9.6 INERTIA

Every load has to be started. In the process, torque has to be supplied to accelerate the load to operating speed and to store in it the kinetic energy associated with its motion. It is entirely possible that the torque required to start the load is greater than the torque required to operate it at steady state.

The inertia of the load does not stand alone as a factor which determines the starting torque that the drive will experience. The motor has a characteristic starting torque and a characteristic inertia of its own. The motor must start the combined inertia of the motor itself, the drive, and the load. The torque seen by the section of the drive will be the starting torque of the motor multiplied by the ratio of the inertia of the load divided by the total combined inertia.

$$T_{input} = T_{motor} \frac{Wk^2_{load}}{Wk^2_{motor} + Wk^2_{load}}$$

Characteristics of Loads

The method of analyzing the drive includes the concept of "reflecting" the load inertia back to the motor. Inertia reflects inversely as the drive ratio squared. That is, if a drive is operating in a reducer ratio of 3, the inertia of the load reflected to the motor is one-ninth of the inertia of the load. Conversely, if the drive is operating as a speed increaser, the inertia reflected to the motor is greater than the actual load inertia by the speedup ratio squared.

The reason for this is that if the drive has to speed up the load, say in the ratio 1:2 compared to the motor, any change in motor speed has to appear as twice the change in drive speed. As the motor starts and, in a fraction of a second, goes from 0 to 100 rpm, the load, in the same fraction of a second, has to go from 0 to 200 rpm. The torque available to do this is only half of the motor torque, because of the speed ratio of the drive. (Input power and output power have to be equal, neglecting losses. Power is torque times speed. If output speed is higher than input speed, the output torque has to be lower than input torque in the same ratio.)

When the mathematics are worked out, it is seen that the drive ratio works against you twice. The speed change of the load is greater than the speed change of the motor by the drive ratio, but the torque available to accomplish the change is lower by the drive ratio. The two are multiplied to give the combined effect, or reflection. The situation works in your favor, of course, when the drive is set up as a speed reducer.

Examples of the use of this information are given in Chapter 12. Inertia is important for starting and stopping considerations and for considerations of speed change during operation. Either of these may turn out to be dominant factors in drive selection and application.

9.7 CHARACTERISTICS OF LOADS

Table 9.3 gives the characteristics of some large loads in terms of the requirements they place on synchronous electric motors. The table is normalized in terms of percent of full-load torque. To be valid in the selection of an electric motor drive, the full-load torque rating of the motor must be at least equal to the full-load torque rating of the load under the operating conditions imposed on the electric motor.

The terms used in the table headings have been described in Section 8.2.5. Briefly recapitulated they are:

> Locked rotor is the torque required to break the load away from a standstill. This takes into account load friction and the load torque-speed characteristic.

TABLE 9.3 Characteristics of Loads

Item No.	Application	Torques in Percent of Motor Full-load Torque			Ratio of Wk^2 of Load to "Normal Wk^2 of Load"
		Locked-rotor	Pull-in	Pull-out	
1	Attrition Mills (for grain processing)—starting unloaded	100	60	175	3–15
2	Ball Mills (for rock and coal)	140	110	175	2–4
3	Ball Mills (for ore)	150	110	175	1.5–4
4	Banbury Mixers	125	125	250	0.2–1
5	Band Mills	40	40	250	50–110
6	Beaters, Standard	125	100	150	3–15
7	Beaters, Breaker	125	100	200	3–15
8	Blowers, Centrifugal—starting with:				
	a. Inlet or discharge valve closed	30	40–60*	150	3–30
	b. Inlet and discharge valve open	30	100	150	3–30
9	Blowers, Positive Displacement, Rotary—by-passed for starting	30	25	150	3–8
10	Bowl Mills (coal pulverizers)—starting unloaded				
	a. Common motor for mill and exhaust fan	90	80	150	5–15
	b. Individual motor for mill	140	50	150	4–10
11	Chippers—starting empty	60	50	250	10–100
12	Compressors, Centrifugal—starting with:				
	a. Inlet or discharge valve closed	30	40–60*	150	3–30
	b. Inlet and discharge valve open	30	100	150	3–30
13	Compressors, Fuller Company				
	a. Starting unloaded (by-pass open)	60	60	150	0.5–2
	b. Starting loaded (by-pass closed)	60	100	150	0.5–2
14	Compressors, Nash-Hytor—starting unloaded	40	60	150	2–4
15	Compressors, Reciprocating—starting unloaded				
	a. Air and gas	30	25	150	0.2–15
	b. Ammonia (discharge pressure 100–250 psi)	30	25	150	0.2–15
	c. Freon	30	40	150	0.2–15
16	Crushers, Bradley-Hercules—starting unloaded	100	100	250	2–4
17	Crushers, Cone—starting unloaded	100	100	250	1–2
18	Crushers, Gyratory—starting unloaded	100	100	250	1–2
19	Crushers, Jaw—starting unloaded	150	100	250	10–50
20	Crushers, Roll—starting unloaded	150	100	250	2–3
21	Defibrators (see beaters, standard)				
22	Disintegrators, Pulp (see beaters, standard)				
23	Edgers	40	40	250	5–10
24	Fans, Centrifugal (except sintering fans)—starting with:				
	a. Inlet or discharge valve closed	30	40–60*	150	5–60
	b. Inlet and discharge valve open	30	100	150	5–60
25	Fans, Centrifugal Sintering—starting with inlet gates closed	40	100	150	5–60
26	Fans, Propeller Type—starting with discharge valve open	30	100	150	5–60
27	Generators, Alternating Current	20	10	150	2–15
28	Generators, Direct Current (except electroplating)				
	a. 150 kW and smaller	20	10	150	2–3
	b. Over 150 kW	20	10	200	2–3
29	Generators, Electroplating	20	10	150	2–3
30	Grinders, Pulp, Single, Long Magazine-type—starting unloaded	50	40	150	2–5
31	Grinders, Pulp, All Except Single, Long Magazine-type—starting unloaded	40	30	150	1–5

(Continued)

Characteristics of Loads

Item No.	Application	Torques in Percent of Motor Full-load Torque			Ratio of Wk^2 of Load to "Normal Wk^2 of Load"
		Locked-rotor	Pull-in	Pull-out	
32	Hammer Mills—starting unloaded	100	80	250	30–60
33	Hydrapulpers, Continuous Type	125	125	150	5–15
34	Jordans (see refiners, conical)				
35	Line Shafts, Flour Mill	175	100	150	5–15
36	Line Shafts, Rubber Mill	125	110	225	0.5–1
37	Plasticators	125	125	250	0.5–1
38	Pulverizers, B&W—starting unloaded				
	a. Common motor for mill and exhaust fan	105	100	175	20–60
	b. Individual motor for mill	175	100	175	4–10
39	Pumps, Axial Flow, Adjustable Blade—starting with:				
	a. Casing dry	5–40†	15	150	0.2–2
	b. Casing filled, blades feathered	5–40†	40	150	0.2–2
40	Pumps, Axial Flow, Fixed Blade—starting with:				
	a. Casing dry	5–40†	15	150	0.2–2
	b. Casing filled, discharge closed	5–40†	175–250*	150	0.2–2
	c. Casing filled, discharge open	5–40†	100	150	0.2–2
41	Pumps, Centrifugal, Francis Impeller—starting with:				
	a. Casing dry	5–40†	15	150	0.2–2
	b. Casing filled, discharge closed	5–40†	60–80*	150	0.2–2
	c. Casing filled, discharge open	5–40†	100	150	0.2–2
42	Pumps, Centrifugal, Radial Impeller—starting with:				
	a. Casing dry	5–40†	15	150	0.2–2
	b. Casing filled, discharge closed	5–40†	40–60*	150	0.2–2
	c. Casing filled, discharge open	5–40†	100	150	0.2–2
43	Pumps, Mixed Flow—starting with:				
	a. Casing dry	5–40†	15	150	0.2–2
	b. Casing filled, discharge closed	5–40†	80–125	150	0.2–2
	c. Casing filled, discharge open	5–40†	100	150	0.2–2
44	Pumps, Reciprocating—starting with:				
	a. Cylinders dry	40	30	150	0.2–15
	b. By-pass open	40	40	150	0.2–15
	c. No by-pass (three cylinder)	150	100	150	0.2–15
45	Refiners, Conical (Jordans, Hydrafiners, Claflins, Mordens)—starting with plug out	50	50–100**	150	2–20
46	Refiners, Disc Type—starting unloaded	50	50	150	1–20
47	Rod mills (for ore grinding)	160	120	175	1.5–4
48	Rolling Mills				
	a. Structural and rail roughing mills	40	30	300–400‡	0.5–1
	b. Structural and rail finishing mills	40	30	250	0.5–1
	c. Plate mills	40	30	300–400‡	0.5–1
	d. Merchant mill trains	60	40	250	0.5–1
	e. Billet, skelp, and sheet bar mills, continuous, with lay-shaft drive	60	40	250	0.5–1
	f. Rod mills, continuous with lay-shaft drive	100	60	250	0.5–1
	g. Hot strip mills, continuous, individual drive roughing stands	50	40	250	0.5–1
	h. Tube piercing and expanding mills	60	40	300–400‡	0.5–1
	i. Tube rolling (plug) mills	60	40	250	0.5–1
	j. Tube reeling mills	60	40	250	0.5–1
	k. Brass and copper roughing mills	50	40	250	0.5–1
	l. Brass and copper finishing mills	150	125	250	0.5–1
49	Rubber Mills, Individual Drive	125	125	250	0.5–1
50	Saws, Band (see band mills)				
51	Saws, Edger (see edgers)				
52	Saws, Trimmer	40	40	250	5–10
53	Tube Mills (see ball mills)				

Table 9.3 (continued)

Item No.	Application	Torques in Percent of Motor Full-load Torque			Ratio of Wk^2 of Load to "Normal Wk^2 of Load"
		Locked-rotor	Pull-in	Pull-out	
54	Vacuum Pumps, Hytor				
	a. With unloader	40	30	150	2–4
	b. Without unloader	60	100	150	2–4
55	Vacuum Pumps, Reciprocating—starting unloaded	40	60	150	0.2–15
56	Wood Hogs	60	50	250	30–100

*The pull-in torque varies with the design and operating conditions. The machinery manufacturer should be consulted.
†For horizontal shaft pumps and vertical shaft pumps having no thrust bearing (entire thrust load caried by the motor), the locked-rotor torque required is usually between 5 and 20 percent, while for vertical shaft machines having their own thrust bearing a locked-rotor torque as high as 40 percent is sometimes required.
‡The pull-out torque varies depending upon the rolling schedule.
**The pull-in torque required varies with the design of the refiner. The machinery manufacturer should be consulted. Furthermore, even though 50 percent pull-in torque is adequate with the plug out, it is sometimes considered desirable to specify 100 percent to cover the possibility that a start will be attempted without complete retraction of the plug.
Recommended Standard 5-20-1938. Authorized Engineering Information 6-24-1949, revised 7-12-1961.
Source: NEMA MG 1-21.87; courtesy of the National Electrical Manufacturers Association.

> Pull-in is the torque imposed by the load just as full speed is reached. The synchronous motor can pull the load into synchronism if its pull-in torque capability equals or exceeds the pull-in load, and if the inertia of the load falls within a certain range for which the pull-in torque rating of the motor is valid.
>
> Pull-out is the torque that tends to unlock the motor from synchronous speed. The motor pull-out torque rating must be at least equal to the torque peaks imposed by the load or the motor may unlock from synchronism. In a way the pull-out torque figure corresponds to the service factor, or shock factor, of the load as estimated in Table 9.2

The right-hand column of Table 9.3 refers to inertia, Wk^2, by normalizing it to a "normal Wk^2 of load." This column can be interpreted as follows:

$$\text{Normal Wk}^2 \text{ of load} = \frac{0.375 \times \text{horsepower}^{1.15}}{(\text{speed}/1000 \text{ rpm})^2}$$

The horsepower and speed of the drive motor will be known. By plugging into this formula, the "normal Wk² of load" can be found. Then by multiplying by the factor in the right-hand column, the inertia of the load can be decoded. These figures have a wide range. They may be useful in taking a first look at a proposed application, but if the design appears that it may not have a large margin for safety in the worst range of the table, the application engineer would do well to obtain a reasonable estimate of the load inertia for the application being considered. This inertia reflects to the drive motor inversely by the square of the speed ratio of the drive, as discussed in Section 9.6.

10

THE DRIVE SPECIFICATION SHEET

When a traction drive is to be specified, certain information must be assembled for presentation to the drive manufacturer or drive distributor, to be used as the basis for recommendation of the most appropriate model in inventory. Each manufacturer offers an application sheet that lists the information required. This chapter is developed in the format of a composite specification sheet with a discussion associated with each item showing how it influences the ultimate recommendation. In conjunction with this list, several items that have to be considered by the user have been added. Figure 10.1 is the composite data sheet that will be used.

Introductory Items

Items 1 to 4 identify you to the manufacturer. It is always helpful if the manufacturer has a point of contact for further information.

Item 5 tells a little about the application just by naming it. For example, by referring back to characteristics of loads (Chapter 9), shock loading and inertia may be checked against what might be considered typical for that kind of application.

Items 6 and 7 are attention-getters if large. Understandably, more attention is given to quotations for drives to be incorporated in ongoing production applications than is given to one-of-a-kind applications. No manufacturer can afford the reputation they would get if their equipment failed to perform in service, even in one-of-a-kind applications. All inquiries should be given a competent technical check.

Input Items

Horsepower, speed, and torque are related by equation (4.2). When the drive is to be a variable-speed one, the usual format is for the output to provide the speed variation for relatively constant input conditions. Therefore, input horsepower (8) is a good baseline against which to begin to examine the drive requirement. Input speed (9) is important because, in conjunction with horsepower it establishes or confirms torque. It will also form the baseline of the life calculation, which will be in the format of millions of stress cycles. Number of stress cycles, divided by speed, and treated by the appropriate constants, will be life, in hours. Drives are usually not sensitive to speed per se. The available drives will all operate at input speeds in the range of commercially available motors and engines.

Torque (10) is probably the most important factor in sizing the drive. It is the factor that requires the contact force, which, in conjunction with the traction coefficient available from the design and the fluid, provides the tangential drive force at the contact. The contact force establishes the stress level. The stress level establishes the life. Torque may be different at the input than at the output. If the drive is a step-up unit (output speed higher than input speed), then the input torque determines the drive loading. If the drive is a step-down unit (output speed lower than input speed), the maximum output torque determines the drive loading. If the drive is adjustable either way, as many drives are, both torque maxima have to be checked. Drives advertised as "constant-horsepower" drives are invariably limited, either by the speed ratio selectable or by the rating itself, to the output torque at the greatest step-down ratio.

Motor rating and type (11) tell a good deal about the application. When a traction drive is driven by a motor, the input torque could be that torque required by the load (factored by the drive gear ratio and efficiency, of course), as long as that torque is not greater than the motor can provide. The analyst has to look at both ends of the drive to see which really provides the limiting criteria. Section 8.2 gives the characteristics of many motors. Whereas the motors are normally thought of in terms of their rating point, some types have a much greater starting torque than their rated running torque and could possibly distress the drive during the starting transient if the load has a high inertia. Most drives have a torque-sensitive loading device which allows them to sustain high torque peaks without gross slip. The preload mechanism has to be considered against the input capability as well as the load capability to make sure that the limiting one has not been overlooked.

Some methods of input coupling (12) provide substantial side load which would have to be supported by the input section. V-belts

Drive Specification Sheet

Company (1)   Individual (4)   Quantity (6)
Address (2)   Application (5)   Estimated Annual Usage (7)
Telephone (3)

Input
  Horsepower (8)
  Speed (9)
  Torque (10)
  Motor rating and type (11)
  Method of input coupling (12)
    Coupling type (12a)
    Gear (12b)
    Belt type (12c)
    Chain (12d)
    Face mount (12e)
    Other (12f)
  Input shaft size (13)
    Metric (13a)
    Inch (13b)
  Direction (14)
    Nonreversing (14a)
    Reversing (14b)

Output
  Speed range required (15)
  Torque at maximum speed (16)
  Torque at minimum speed (17)
  Is an output gearbox desired? (18)
    Ratio (18a)
    Shaft arrangement (18b)
  Method of output coupling (19)
    Coupling type (19a)
    Gear (19b)
    Belt (19c)
    Chain (19d)
    Face mount (19e)
    Other (19f)
  Output shaft size (20)
    Metric (20a)
    Inch (20b)

Adjustment
  Local (21)
    Handwheel (21a)
    Vernier (21b)
    Indicator (21c)
    Which face (21d)
    Number of turns (21e)
  Remote (22)
    Mechanical (22a)
    Electrical (22b)
    Pneumatic (22c)
    Hydraulic (22d)
    Indicator (22e)
    Which face (22f)
    Number of turns (22g)

Mounting
  Horizontal (23)
    Feet down (23a)
    Feet up (23b)
  Vertical (24)
    Output shaft facing up (24a)
    Output shaft facing down (24b)

Load
  Inertia (Wk^2) (25)
  Shock (26)
    Light (26a)
    Uniform (26b)
    Moderate (26c)
    Severe (26d)
    Very severe (26e)
  Starts per hour (27)
  Reversals per hour (28)
  Speed changes per hour (29)
    Full range (29a)
    Narrow range (29b)
  Speed-holding precision (30)
    Tight, 0.1% (30a)
    Moderate, ½% (30b)
    Not critical, 2% or more (30c)

Drive Specification Sheet   249

[Load]
  Speed adjustment (31)
    Only while running (31a)
    Adjustment at rest required (31b)
    Does the adjustment go through zero speed? (31c)
    Repeatability (31d)
  Spectrum (32)
    Percent of time spent at each load step (32a)
    Can the load overhaul? (32b)

Operating Conditions
  Temperature (33)
  Humidity (34)
  Hazards (35)
    Dust (35a)
    Lint (35b)
    Chemicals (35c)
    Fumes/explosion (35d)
  Hours per day (36)

Access
  To remove motor (37)
  To inspect lubricant (38)
  To remove drive (39)

Power Source
  Electric motor (40)
    Voltage (40a)
    Number of phases (40b)
    Frequency (40c)
    Type (40d)
  Other (41)

FIG. 10.1 Drive specification sheet.

and flat belts (12c) provide torque by the difference in tension between the pulling and return sides of the belt. If the return side is slack, as is often the case, the force in the pulling side is simply the input torque divided by the radius of the input pulley. If the belt tension is maintained by a dancer or a fixed idler, the return side may not be slack and the side load could be higher. Spring-loaded motor mounts which are pivoted so that the motor torque increases belt tension may be set up to increase tension much more than the torque requires. Most drives do not have a support outboard of the pulley. Therefore, side loads created by input or output loading are often called overhung loads.

A formula frequently used to evaluate overhung load on a drive is

$$\text{Overhung load} = \frac{126{,}050 \times \text{hp} \times C_f \times L_f}{\text{rpm} \times \text{dia}_{\text{pulley}}}$$

where

C_f = 1.1 for chain drive
  1.25 for gear drive
  1.5 for V-belt drive
  2.5 for flat-belt drive
L_f = 1.1
H_p = drive horsepower
rpm = drive speed, rpm
dia_{pulley} = pulley diameter, in.

Note that chain drives (12d) provide the most favorable overhung loading of the side-loading types. This is because the return side is almost always slack. Chain drives provide a periodic force as the rollers engage each sprocket tooth due to a phenomenon called chordal acceleration.

Gearing (12b), which provides the next most favorable overhung loading, provides a side load not only due to the tangential force applied at the pitch radius of the gear, but also due to a separating force which is a function of the tangential force and the pressure angle of the gearing. A pressure-angle gearing of 20° provides a separating force equal to 36% of the tangential force. A pressure-angle gearing of $14\frac{1}{2}°$ provides a separating force equal to 26% of the tangential force. The only form of gearing that does not generate any side force is the planetary, in which the multiple planets balance the forces on the sun and ring gears to provide essentially pure torque. Input gearing could be installed in a structural housing which reacts all side loads internally and is coupled to the drive by one of the methods that provide pure torque.

In-line shaft couplings (12a) provide pure torque when properly applied. There are many formats of commercial couplings available which do a good job. The type selected should, ideally, provide for any offsets and angular misalignments encountered in the installation.

Face mounting (12e) creates a compact assembly and takes up most of the tolerance problems associated with alignment. The motor is usually a NEMA C flange (tapped flange) or D flange (drilled flange) type (see Section 8.2.3 for descriptions), or the comparable IEC versions. The drive has a complementary mating geometry. Sometimes the motor mounts on the drive (the drive has the mounting legs). Installations have also been seen in which the drive mounts on the motor. It is important that the motor and the drive be physically compatible. The drive flange must mate with the flange on the motor selected. If the drive is of the IEC standard type, an IEC standard motor must be selected. Some drive manufacturers furnish adapter flanges so that their metric-style mountings can be used with NEMA standard motors.

Drive Specification Sheet

Input shaft size (13) usually is not specified as a stand-alone requirement. It is characteristic of the drive which the manufacturer selects. It must be compatible with the motor that is going to be used. By specifying the motor rating and type (11) and the intended method of input coupling (12) the user has provided all the necessary information for this check. The drive might be from a metric series or from an inch series. Adapters may be recommended to provide the necessary compatibility.

Direction of rotation (14) may or may not be important. Many drives on the market can be operated in either direction; some cannot. Drives such as Wedgtrac and a Shimpo ring-cone variator can be rotated in only one direction, either clockwise or counterclockwise, which must be specified at the time of ordering. A feature of the Wedgtrac drive is that if the drive is reversed, the rollers will disengage. Specify the direction of rotation to be sure, and specify whether the direction of the input rotation is subject to reversals of direction. Direction of rotation is one simple technical concept which can be the subject of enormous confusion because one often assumes that the other fellow is taking the same point of view as one's own. Avoid using terms such as "front" or "back." Drives should be specified as "rotation clockwise looking at the input shaft" (or counterclockwise as the case may be). The recent trend to digital clocks does not help the case, but at least we can be clear about where we are standing when we declare the direction.

Output Items

Output speed range (15) is an important parameter. Some manufacturers feature a 6:1 range or a 9:1 range in which the range is the maximum output speed (drive in step-up mode) divided by the minimum output speed (drive in step-down mode). The midpoint of the range may or may not necessarily be balanced around the 1:1 speed ratio. For example, when a 6:1 range is offered, it may be that the drive can step its output speed down to one-third of the input speed, or up, to two times the input speed. Some 9:1 drives go from one-third of input speed to three times input speed at the output. Since torque sizes the drive structurally, the lowest output speed needed generally establishes the frame size of the drive.

Some manufacturers offer a very narrow speed range drive, but within this range the output speed can be set with great resolution and maintained with great precision. Some manufacturers can select the range being offered by how the internal parts are assembled. When very high output speeds are to be experienced, extra balancing steps are taken and the grade of steel of certain parts may be changed.

Output torque at maximum speed (16) defines one possible horsepower maximum which will be checked against input horsepower (8). Output torque at minimum speed (17) is also a check on input horsepower (8). The output torque at minimum speed probably sizes the drive from a structural standpoint. The horsepower figures when used in conjunction with the efficiency of the drive establish the heat losses in the drive. These are examined against the ratings to be sure that the drive will not be overheated in service.

An output gearbox (18) can be used to change the orientation of the output shaft by offsetting it parallel to the original direction, or by turning it through 90° (or any other desired angle). More important, the output gearbox can shift the nominal speed up or down so that the traction drive need only be concerned with speed ratio with respect to the nominal output speed. The traction drive does not have to offset the nominal speed with respect to the motor speed as well as provide a speed range around the new nominal. Most traction drive manufacturers offer a wide range of standard gearboxes as accessories to their drives. If the midrange of the required output speed variation is to be substantially different from the input speed, a gearbox would be a good accessory to consider. If the output speed is to be lower than the input, the gearbox could provide extra torque which does not have to pass through the traction elements. Conversely, if the gearbox is to be a speed-up ratio, the traction drive is spared some of the output revolutions, which, when added up, determine its life.

The method of output coupling (19) raises the same considerations listed under input coupling (12). Since loads tend not to be standardized, it is not probable that the drive manufacturer has a face-mount option readily available.

The output shaft size (20) raises the same considerations as mentioned under input shaft (13). Usually, the torque requirement sizes the drive and the manufacturer has a standard shaft size for each size of drive. There will be different models for metric series or inch series drives, and in some cases one can be converted to the other with adapters.

Adjustment Items

If the drive is to be an adjustable output speed device, the question immediately arises as to where to locate the adjustment. The adjustment may be local (21), that is, immediately on the drive, or it may be remote (22), mounted on a panel nearby or slaved to a remote control. In either case the access point to the drive must be chosen. It would be disastrous to discover, upon drive installation, that access to the adjustment wheel is blocked by local structure or by other equipment. Most drive manufacturers offer two or three adjustment access

Drive Specification Sheet

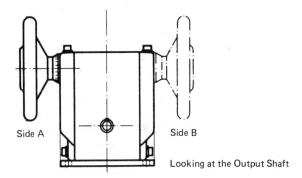

Looking at the Output Shaft

FIG. 10.2 Schematics of handwheel position selections. (Courtesy of Plessey Dynamics Corporation, Hillside, N.J.)

positions (left side, right side, top). Figure 10.2 shows one manufacturer's adjustment selection.

The adjustment input device is usually offered in several forms. A handwheel (21a) is common. If the handwheel must make many turns to accomplish the full adjustment range, the position of the drive ratio adjustment is not determinable by just looking at the wheel. An indicator or counter will be required.

In some drives the adjustment may be made by a vernier screw in which marketings on the edge of the adjusting knob or collar may be referenced to scribed lines on the housing. This method will indicate the adjustment position with considerable resolution.

The number of turns required to accomplish the full adjustment range (21e) can be specified. Many manufacturers offer an auxiliary gear box which attaches to the adjusting shaft which can multiply the required adjustment motion by factors of from 6 to 50. The operator may traverse the adjustment range quickly at about 2 revolutions per second, then slow down to 1 revolution per second near the end. If the maximum precision gearbox with 50:1 ratio is selected, and the drive itself has an adjustment range which requires 30 revolutions from one extreme to the other, when adjusted at the rate of 2 revolutions per second it might take the following time to traverse the adjustment range:

$$T = 30 \times \frac{50}{2}$$
$$= 750 \text{ sec}$$
$$= 12.5 \text{ min}$$

The time to accomplish the required adjustment has to be thought through before the adjustment turns are specified.

If the adjustment is to be remote, most manufacturers have several options available. The simplest would be a flexible cable connecting a handwheel on a control panel with the adjustment shaft on the drive. The turns indicator would be mounted next to the handwheel. If the handwheel is to control some function (usually speed), it would be a good idea to have a remote speed readout, either analog or digital, on the panel alongside the handwheel for the operator's information.

Remote adjustment (22) lends itself to servo control. A servo system can be set up to control the drive output in terms of speed or some other function of speed. The servo would require an input or a sensor which tells it the value of the variable it is controlling. The output could be connected to the drive adjustment shaft. Servos could be electric (22b) and could drive the adjustment shaft with ac or dc control motors. Servos could also be pneumatic (22c) or hydraulic (22d), the choice being dependent on the equipment designer and what is available in the marketplace. It would not be unusual to use a pneumatic servo in a situation where there could be danger of explosion in the presence of a brush spark or if an accumulation of lint might block the normal cooling mechanism.

Mounting Items

Mounting can be horizontal (23) or vertical (24). If the shaft is to be horizontal, the mounting feet can be down (23a) or, if the unit is to be mounted on the underside of a surface, up (23b). If the shaft is to be vertical, the unit could be mounted with the output shaft facing up (24a) or down (24b). The manufacturer will need to know about the intended mounting so that the lubricant fill port and sight glass will end up in the right place. Figure 10.3 shows some possible mounting arrangements.

Load Items

Inertia (25) is a very important characteristic in the sizing of the drive. Whenever speed changes, the load either gives up energy or acquires it. The largest transfer of energy usually occurs during starts. The motor, coupled to the load through the drive, is thrown across the line. The motor develops its locked-rotor torque. This torque is used to accelerate the motor itself and the load. In most cases the locked-rotor torque is larger than the normal operating torque. Table 8.9 lists this ratio for NEMA design motors.

The torque transferred to the drive is related to the motor locked-rotor torque, the motor inertia, and the load inertia reflected to the motor shaft, in accordance with the following equation:

$$T_{drive} = \frac{(T_{motor} - T_{load}) Wk^2_{load\ refl}}{Wk^2_{motor} + Wk^2_{load\ refl}} + T_{load} \qquad (10.1)$$

Drive Specification Sheet

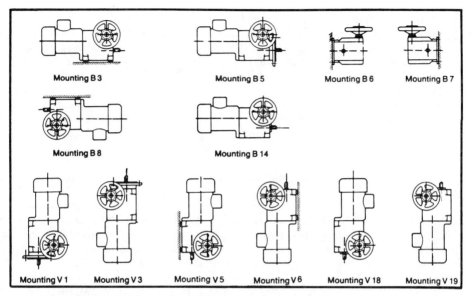

FIG. 10.3 Schematics of mounting position selections. (Courtesy of Plessey Dynamics Corporation, Hillside, N.J.)

where

T_{drive} = torque transmitted to the drive, lb ft
T_{motor} = motor locked-rotor torque, lb ft
T_{load} = load friction or steady-state opposing torque component, lb ft
$Wk^2_{load\ refl}$ = load inertia reflected to the drive input shaft in accordance with equation (10.2), lb ft^2
Wk^2_{motor} = motor rotor inertia, lb ft^2

Load inertia must be reflected to the drive input shaft in order for expression (10.1) to be valid. This may be done in accordance with the following relationship:

$$Wk^2_{reflected\ to\ drive\ input} = Wk^2_{load} \left(\frac{rpm_{load}}{rpm_{drive\ input}} \right)^2 \quad (10.2)$$

where

Wk^2_{load} = inertia of the load, lb ft^2
rpm_{load} = speed at which the load would turn, based on the drive ratio and the speed at which the motor would turn
rpm_{motor} = motor speed

Section 9.5 describes the reason for this relationship.

The inertia of the load is calculated by the techniques of Chapter 4 and by the use of Table 4.2. If the load consists of several parts, each of significant inertia, which themselves are turning at different speeds, equation (10.2) may be used to reflect the various parts of the load separately.

Figure 8.11 gives the inertias of some drive motors. There are so many different motors available that it is not practical to try to tabulate them all or to try to identify a particular motor to one of the tabulated values. Figure 8.11 may be used for rough calculations. Then, when a drive selection is about to be confirmed, the motor manufacturer should be called and asked for the inertia of the particular motor of interest.

Shock (26) is a ratio of recurrent peak torque to average torque. The implication in the use of the term "shock" rather than "maximum torque" is that the peaks are very brief, rising sharply and decaying rapidly. Therefore, the peaks do not contain enough energy to affect the motor size. Each peak impresses a stress cycle on the drive. Fatigue, which may be expressed in terms of number of stress cycles to failure, is definitely a life determinant factor. Shock is difficult to characterize except in general terms such as light (26a), uniform (26b), moderate (26c), severe (26d), or very severe (26e). There is a general consensus that, depending on the severity of the shock characterization, the drive load rating should be increased by a factor to accommodate it. Table 9.1 describes the shock conditions normally associated with certain loads in terms of a "service factor" to be applied to the nominal load rating of the drive. Table 9.2 decodes the terms noted above in conjunction with the number of starts per day (each of which is a shock of sorts since the motor starting torque available to accelerate the drive is greater than its nominal running torque).

Starts per hour (27) and reversals per hour (28) are treated in the same manner as shock described above. Each start or each reversal puts the entire drive through a load peak which has to be accounted for in the ultimate assessment of its life. In Table 9.2 the effect of number of starts can be separated from the effect of the various shock categories by recognizing that the first category, light service, has a recommended service factor of 1.0 for 1 to 10 starts per day. Light service is essentially shock-free operation, so the service factors associated with the larger number of starts per day for light service are in recognition of the number of starts on the required drive rating.

The manufacturer will want to assess speed changes per hour (29) to see how much usage the speed-adjust mechanism will experience. In some drives the elements are free to creep with respect to each other in a way that would distribute any wear over the entire surface of the element. Drives that use free balls or ball and disc are notable examples. Even if the speed ratio is never changed in the life of the drive, the balls will experience uniform surface wear.

For drives in which all elements are constrained to rotate on fixed axles, it is conceivable that only a small band of surface will use up all its life potential if the speed ratio is never changed, leaving some surface areas unused. The drive manufacturer needs to appreciate this aspect of intended usage to rate the drive for your application.

It is not necessarily bad practice to leave a drive in a fixed ratio for long periods of usage. Some drives are designed as fixed-ratio drives. Others, as noted above, distribute the wear even in fixed-ratio modes. The intended usage simply has to be known to have the drive properly rated.

Traction drives can never be used for absolutely fixed ratios, for they do creep. In designing the drives it is possible to trade off certain factors that will result in greater speed-holding precision (30) with load changes at the expense of other features. It is easy to see that a drive with a set of balanced opposed rollers will deflect less (and hold speed setting better) than one with a single-sided contact upon increase in load, but it could cost and weigh more. Most loads do not require a great degree of speed-holding precision. The induction motors which drive them slip (with respect to the synchronous line frequency) 3 to 4%, and this will vary with load.

It is well to appreciate the speed-holding capability that will be required from the drive and motor combination, and make sure that the drive being considered is capable of meeting the proposed criterion. If there are cost differences between various drive geometries, make sure that speed-holding requirements are neither overstated nor understated so that you get what you need, and do not pay for unneeded precision.

In one application, courtesy of F.U. (London) Ltd., a wallpaper manufacturer used two kinds of rollers. One embossed a pattern and the other coated the background color. The embossing roller had to be changed every time a different pattern was to be run. The embossing roller diameter depended on the pattern repeat distance. Whenever an embossing roller of a different diameter was loaded, it was customary to change the coating roller to one of the same diameter as the embossing roller so that both could be driven at the same surface speed by a common drive. It proved economical to change the coating roller drive to a variable-ratio traction drive so that the surface speed of the coating roller could be matched to the surface speed of the embossing roller without changing the coating roller every time the embossing roller was changed. This drive did not have to be very precise because the uniform background coating did not have any phase or registry requirements with respect to the pattern roller. Depending on the design of the coating process, the roller might be allowed to slip a small amount with respect to the web, or if it clamped to the web, it might have to be servo controlled as a function of web tension to avoid tearing the web or letting it go slack due to cumulative error.

Speed-holding precision might be characterized as tight (30a), moderate (30b), or not critical (30c) for a range of no load to full load at any ratio setting.

It is important to recognize that some drive formats permit speed adjustment (31) when the drive is at rest. Others require the drive to be in motion before the adjustment mechanism can be moved. If the speed adjustment need be made while the drive is at rest, this must be highlighted on the drive specification sheet. Drives that may be adjusted while at rest include Arter, Contraves, Floyd, Graham planetary ring cone, Hans Heynau, P.I.V. Antreib Werner Reimers, Shimpo ring and movable cone, and Simplana. The last-mentioned drive may be adjusted from a higher to a lower speed only while at rest.

Some drives are equipped with differential gearing arranged so that the output speed can go through zero to reverse at full-rated torque. Such drives can be run at zero output speed continuously without damage. Differential gearing is available as an option on Arter, Chery, Contraves, Floyd, Graham, Heynau, Shimpo, and Vadetec drives. P.I.V. Antreib Werner Reimers and Graham ball disc drives can go to zero output speed without planetary gearing.

If the drive does not have special provision for doing so, it is not good practice to operate at zero speed for any length of time without differential gearing. This induces a 100% spin condition which could be damaging to the contacts.

Adjustment repeatability (31d) may be an important factor in some applications where the ratio has to be preset while the drive is not operating. It should be less important if the ratio adjustment is servo operated or if a system operator uses the adjustment to achieve a certain operating condition. If adjustment repeatability is an important factor in your application, specify it so that the requirement may be evaluated and confirmed.

Spectrum (32) refers to the anticipated distribution of load torque over a typical duty cycle of the drive. Most drives have loading cams which increase contact pressure in proportion to input or output torque. As discussed in Chapter 7, life is inversely proportional to stress raised to the tenth power. Fortunately, stress is only a square root function of load, so the load/life relationship is sensitive only to the cube (approximately) of load.

The techniques of Chapter 7 allow life to be predicted based on a root-mean cubic analysis in which the load steps are listed together

Drive Specification Sheet 259

with the percentage of operating life associated with each. The analyst sums the product of the load cubed times the percentage life for each entry, then takes the cube root of the result for the mean load to be used for life analysis. If a load spectrum is absent, the maximum load has to be assumed to be present all the time.

Overhauling load (32b) is a special case which may have to be addressed. Not all drives can handle loads which at times want to run faster than the drive and have to be retarded by the drive. A hoist is a simple example. When the load is being lowered, it attempts to lower faster than the drive that must retard it. The drive must provide a torque in the direction of raising the load, while turning in the direction to lower the load.

Operating Conditions

Most mechanical drives have a wide range of operating conditions (33-36) under which they will perform satisfactorily. Nonetheless, knowledge of the operating conditions is essential to proper selection of the lubricant grade and to the recommended operating procedures. Drives that have very low operating temperature may be rated with a lower-viscosity lubricant than if they were to operate at normal temperature or at high temperature. Low-temperature drives may require a warm-up period before accepting maximum loading.

Operation under hazardous conditions (35) may be one of the criteria for selecting a traction drive over an electronic variable-speed drive. One of the illustrations in Chapter 12 is a drive in a knitting mill. The drive had to be installed in a location in which it could be completely covered with lint without danger of fire or explosion or of damage to the drive. This section of the checklist reminds the rater and the user of the conditions under which the drive is expected to operate satisfactorily.

Access

Most of the elements on this list (37-39) are really reminders to the user not to build the drive in such a way that it cannot easily be inspected or serviced. Emphasis is placed on the word "easily," especially in regard to lubricant level and top-off, because what is not done easily often is not done at all. Some drives may have optional locations for the lubricant sight glass which might better suit your intended application.

Power Source

If the power source is an electric induction motor, many of its characteristics may be inferred from the information in Chapter 8. If the motor is to be furnished by the supplier as part of the drive, it is important to specify the voltage, number of phases, and frequency (40).

Standard voltages (40a) available in the United States include 115 (110-120), 208, 230 (220-440), 460 (440-480). The 115 V service is usually used for small, single-phase applications. Larger, integral-horsepower single-phase motors would probably use 208 or 230 V. Most three-phase motors utilize 208, 230, or 460 V unless they are very large (40b). It would hardly do to have a drive wired for 230 V arrive for use in a 460-V electrical system.

It is important to state frequency (40c) when purchasing drives manufactured in Europe for use in the United States or vice versa. Standard power line frequency in Europe is 50 Hz. European induction motors are designed to run at 1500 rpm (4-pole). Corresponding American (60-Hz) motors are designed to run at 1800 rpm. 115-V motors are not available in Europe. Preferred power voltage is 240 for single-phase motors and 415 for polyphase motors. If the user is purchasing a drive to operate with a motor which has already been selected, then the drive manufacturer will need to know the above information to establish the torque speed input characteristic.

Traction drives are not necessarily powered by electric alternating current induction motors. They could be powered by direct current motors, hydraulic motors, gasoline engines, Diesel engines, turbines, screw expanders (40d). This book concentrates on the characteristics of the induction motor, since these are ubiquitous and well standardized. The user can be guided as to which characteristics to select and which questions to ask. The word "other" in the check list (41) reminds the user that if a "standard" power source is not being used, then it is necessary to work with the drive manufacturer and furnish the corresponding torque speed characteristics and installation information for the source to be used.

11
ECONOMICS

This chapter on economics is included to define the terms that are encountered when the cost aspects of traction drives are evaluated, and to show how they relate to the selection process.

11.1 CAPITAL EQUIPMENT

Although capital equipment can exist in many forms, from the point of view of this text capital equipment may be defined as equipment that has a useful life of more than 1 year. When the cost of a solution to a traction drive application is being considered, the cost of the drive itself will be the capital equipment cost.

11.1.1 Basis

This is a term describing the value of a piece of capital equipment as it stands on the books of the company. When a traction drive is purchased it goes on the books of the company at the purchase price (including installation). Later the basis will be adjusted for depreciation, as described in Section 11.2.

11.2 DEPRECIATION

When a company purchases capital equipment such as a traction drive for its own use, it is not allowed to deduct the full cost of the drive from its income on its corporate tax return for the year in which the drive was purchased. Only part of the difference between the acquisition cost and the estimated salvage value may be offset against the company's income each year as a cost of doing business. This process is called depreciation.

The concept of depreciation relates to the wear and tear suffered by capital equipment. The treatment of it relates wholly to the way in which the tax laws are written. The tax laws establish minimum depreciation periods for various kinds of capital assets. "Equipment" such as traction drives may not be depreciated in less than 5 years unless the company can prove that a shorter lifetime is applicable.

There are four depreciation methods sanctioned by the tax code: (1) straight-line depreciation, (2) sum-of-the-years'-digits method, (3) declining-balance method, and (4) accelerated cost recovery system. In this chapter all illustrations will be made on the basis of the straight-line method. Each of the other methods has its place in special situations.

11.2.1 Straight-Line Depreciation

In this method the difference between the original basis, or value on the books of the company, and an expected scrap or salvage value, is divided into equal parts. If the equipment has a 5-year depreciation life, there would be five equal parts. The basis will be reduced by one-fifth of the difference between the original value and the scrap value each year. Table 11.1 shows how this works.

The scrap value is an accounting artifact. It is frequently set at 10% of the original value, or at zero. The only thing that happens if the estimate was not accurate is that the company has to account

TABLE 11.1 Straight-Line Depreciation for 5-Year Life

Year	Percent depreciation[a]	Adjusted basis[b] (%)	Adjusted basis[c] (%)
1	20	80	82
2	20	60	64
3	20	40	46
4	20	20	28
5	20	0	10

[a]The percentage depreciation is a percentage of the original cost or basis minus the estimated salvage or scrap value
[b]This column is based on an estimated scrap value of zero.
[c]This column shows the computation on the assumption that the salvage or scrap value is 10% of the original basis.

Discounted Cash Flow Analysis

for a capital gain (or loss) if it sells the equipment for more (or less) than the scrap value that had been estimated.

11.3 EXPENSE

Expense is any cost that the tax laws permit to be written off against income in the year in which it occurs. When a traction drive is operated, the electricity it uses is an expense item. The cost of maintenance of the drive would also be an expense item.

11.4 DISCOUNTED CASH FLOW ANALYSIS

Accountants consider whether a candidate capital investment should or should not be made on the basis of a net cash flow analysis which takes into account the tax status of the company. For any one year a net cash flow is calculated based on the estimated change in the following factors:

Revenues, ΔR
Operating costs, ΔO
Depreciation, ΔD

as a result of having made the investment. That is, revenue is estimated as though the capital investment had not been made. Then it is estimated as though the investment had been made, and the difference is ΔR. A similar set of estimates yields ΔO and ΔD.

The net cash flow (NCF) for that year is (Ref. 1, eq. 8.2)

$$NCF = (\Delta R - \Delta O - \Delta D)(1 - t) + \Delta D \qquad (11.1)$$

where t is the marginal tax bracket, that is, the tax bracket which applies to the last dollar earned, expressed as a decimal. Notice that the change in depreciation, ΔD, appears twice. It is operated on by the tax rate because it is a tax deduction, but it is put back at the end of the equation because depreciation, although not taken out of pocket each year, is a very real cost to be charged against investment.

The capital investment, in this case the traction drive, will have been assigned a depreciation life in accordance with Section 11.2. The net cash flow (NCF) is calculated separately for each year of the depreciation period. The present value of each of these future cash flows is calculated by the rules of compound interest. The net present value (NPV) of the proposed investment is the sum of each of these terms, including the purchase cost, as follows:

$$\text{NPV} = -(\text{acquisition cost}) + \frac{\text{NCF}_1}{1+i} + \frac{\text{NCF}_2}{(1+i)^2} + \frac{\text{NCF}_3}{(1+i)^3}$$
$$+ \frac{\text{NCF}}{(1+i)^4} + \frac{\text{NCF}}{(1+i)^5} \qquad (11.2)$$

where i is the interest rate, as a decimal. The number of NCF terms in the equation is the same as the number of years of depreciation life.

Presumably, the sum of the NCF terms should be great enough to offset the acquisition cost, so that the NPV is a positive number. If NPV for a given proposed investment turns out to be negative, the accountants would recommend against making the investment.

Most of the time engineers are faced with the question of which of several alternatives is the best. The decision that a drive is needed will have been made at the outset. The technique for comparing alternatives is to do the NPV calculation for each one and compare them. The candidate with the most positive (least negative) NPV is the best choice.

When comparing traction drives for a given application we need not be concerned with the ΔR terms. We have already decided that we need a drive, and the only question is which one. Since we do not have a traction drive now, the Δ terms may be replaced with the estimated annual operating and depreciation costs of the drive. Rewriting equation (11.1) in this form, we have

$$\text{NCF} = (-O - D)(1 - t) + D \qquad (11.3)$$

or

$$\text{NCF} = -O(1 - t) + Dt$$

11.5 EVALUATION OF ALTERNATIVES

Suppose that two drive proposals are being compared. The original problem statement called for a 2-hp load to be driven at an output speed of from 750 to 1500 rpm using a 1750-rpm motor which was already available. The proposals are as follows:

Proposal A: a drive sized for the highest torque output, which would be seen at 2 hp, 750 rpm. Output speed is adjustable from 750 to 1500 rpm with 1750-rpm input. Price: $2655.
Proposal B: a smaller drive with a step-down gearbox as part of the drive. The step-down box has a 3:1 ratio. Drive output speed is 2250 to 4500 rpm to achieve the required 750 to 1500 rpm at the output of the gearbox. The drive alone

Evaluation of Alternatives 265

would actually be operating in step-up mode with respect to the 1750-rpm input speed. Price: $2059.

The following additional parameters affect the problem:

1. The company is in the 48% tax bracket.
2. The interest rate is 18%.
3. The average electric rate is $0.087 per kilowatt-hour.

Assume that, with its added gearbox and its ratio 0.389 (step up), the less expensive drive is 80% efficient. Assume that the more expensive drive, operating in step-down mode at a ratio of 2.33, is 90% efficient.

The question of comparative life is not addressed here. Chapter 7 shows that life is generally inversely proportional to the cube of load. If the same drive is run faster to reduce its torque load at a given horsepower, the gain in life due to the reduction of torque is greater than the reduction in life due to the greater number of revolutions the drive will have to make. Therefore, proposal B is a valid approach to reducing the cost of a drive. In this analysis, approach B was used to allow the proposal of a smaller frame drive. The financial aspects of proposals A and B are examined assuming that both proposals reflect the same life. This is taken to be 5 years, the same as the allowable depreciation period.

Proposal A

Acquisition cost

A = price plus transportation cost plus installation cost; assume for the purpose of simplicity that all of these costs are covered by the one figure given

= $2655.00

Energy cost. From Table 4.7 we convert 2 hp to kilowatts:

$$\frac{2 \text{ hp}}{1.34102 \text{ hp/kW}} = 1.492 \text{ kW}$$

Given that the drive is 90% efficient and assuming that the motor is 85% efficient, the energy cost for operating the drive for 1 hr would be

$$E = \frac{1.492 \text{ kW} \times \$0.087}{0.90 \times 0.85}$$

= $0.169678

Assuming that the drive will be operated for 2000 hr per year, the annual energy cost would be

$$E = 2000 \times \$0.169675$$
$$= \$339.35 \text{ per year}$$

Maintenance cost. We have no basis, at least from the problem statement as given, to estimate the maintenance cost. Just to illustrate where maintenance cost is treated in this type of analysis, assume that it will be about 5% of the price of the drive, each year. For proposal A:

$$M = 0.05 \times \$2655$$
$$= \$132.75 \text{ per year}$$

The terms to be used in equation (11.3) are:

Operating cost per year, $O = E + M$
$$= \$339.35 + \$132.75$$
$$= \$472.10$$

Depreciation cost per year, $D = \dfrac{\text{acquisition cost}}{5 \text{ years}}$
$$= \dfrac{\$2655}{5}$$
$$= \$531.00 \text{ per year}$$

$$NCF = -O(1 - t) + Dt$$
$$= -\$472.10(0.52) + \$531(0.48)$$
$$= -\$245.49 + \$254.88$$
$$= \$9.39$$

When the substitution is made in equation (11.2), the result is

$$NPV = -A + \dfrac{NCF_1}{1+i} + \dfrac{NCF_2}{(1+i)^2} + \dfrac{NCF_3}{(1+i)^3} + \dfrac{NCF_4}{(1+i)^4} + \dfrac{NCF_5}{(1+i)^5}$$

$$= -2655 + \dfrac{9.39}{1.18} + \dfrac{9.39}{1.18^2} + \dfrac{9.39}{1.18^3} + \dfrac{9.39}{1.18^4} + \dfrac{9.39}{1.18^5}$$

$$= -2655 + \dfrac{9.39}{1.18} + \dfrac{9.39}{1.3924} + \dfrac{9.39}{1.6130} + \dfrac{9.39}{1.9388} + \dfrac{9.39}{2.2878}$$

$$= -2655 + 29.36$$

$$= -2625.64$$

Evaluation of Alternatives

Tables have been prepared which spare the tedium of summing the last five steps of the equation. These steps represent the present value of equal sums to be spent in future years. When equal payments are to be made in this way they are called annuities. Looking in Table 11.2 in the group "for interest rate of 0.18 per annum" in the column headed "present value of an annuity of one dollar" for 5 years, one will read 3.1272. Note that $3.1272 \times 9.39 = 29.36$.

Proposal B

Acquisition cost

$A = \$2059$

Energy cost

$$E = \frac{1.492 \times \$0.087 \times 2000}{0.80 \times 0.85}$$

$= \$381.78$ per year

Maintenance cost

$M = 0.05 \times \$2059$

$= \$102.95$

Operating cost

$O = E + M$

$= 381.78 + 102.95$

$= \$484.73$

Depreciation cost

$$D = \frac{\$2059}{5}$$

$= \$411.80$

$NCF = -O(1 - t) + Dt$

$ = -484.73(0.52) + 411.80(0.48)$

$ = -54.40$

$NPV = -A + NCF \times$ (present value of an annuity of one dollar)

$ = -2059 + (-54.40)(3.1272)$

$ = -2229.12$

TABLE 11.2 Seven-Year Compound Interest and Annuity Table

YEAR	AMOUNT OF ONE DOLLAR AT COMPOUND INTEREST	PRESENT VALUE OF AN ANNUITY OF ONE DOLLAR	PRESENT VALUE OF ONE DOLLAR	YEAR	AMOUNT OF ONE DOLLAR AT COMPOUND INTEREST	PRESENT VALUE OF AN ANNUITY OF ONE DOLLAR	PRESENT VALUE OF ONE DOLLAR
FOR INTEREST RATE 0.05 PER ANNUM				FOR INTEREST RATE 0.1 PER ANNUM			
1	1.0500	.9524	.9524	1	1.1000	.9091	.9091
2	1.1025	1.8594	.9070	2	1.2100	1.7355	.8264
3	1.1576	2.7232	.8638	3	1.3310	2.4869	.7513
4	1.2155	3.5460	.8227	4	1.4641	3.1699	.6830
5	1.2763	4.3295	.7835	5	1.6105	3.7908	.6209
6	1.3401	5.0757	.7462	6	1.7716	4.3553	.5645
7	1.4071	5.7864	.7107	7	1.9487	4.8684	.5132
FOR INTEREST RATE 0.06 PER ANNUM				FOR INTEREST RATE 0.11 PER ANNUM			
1	1.0600	.9434	.9434	1	1.1100	.9009	.9009
2	1.1236	1.8334	.8900	2	1.2321	1.7125	.8116
3	1.1910	2.6730	.8396	3	1.3676	2.4437	.7312
4	1.2625	3.4651	.7921	4	1.5181	3.1024	.6587
5	1.3382	4.2124	.7473	5	1.6851	3.6959	.5935
6	1.4185	4.9173	.7050	6	1.8704	4.2305	.5346
7	1.5036	5.5824	.6651	7	2.0762	4.7122	.4817
FOR INTEREST RATE 0.07 PER ANNUM				FOR INTEREST RATE 0.12 PER ANNUM			
1	1.0700	.9346	.9346	1	1.1200	.8929	.8929
2	1.1449	1.8080	.8734	2	1.2544	1.6901	.7972
3	1.2250	2.6243	.8163	3	1.4049	2.4018	.7118
4	1.3108	3.3872	.7629	4	1.5735	3.0373	.6355
5	1.4026	4.1002	.7130	5	1.7623	3.6048	.5674
6	1.5007	4.7665	.6663	6	1.9738	4.1114	.5066
7	1.6058	5.3893	.6227	7	2.2107	4.5638	.4523
FOR INTEREST RATE 0.08 PER ANNUM				FOR INTEREST RATE 0.13 PER ANNUM			
1	1.0800	.9259	.9259	1	1.1300	.8850	.8850
2	1.1664	1.7833	.8573	2	1.2769	1.6681	.7831
3	1.2597	2.5771	.7938	3	1.4429	2.3612	.6931
4	1.3605	3.3121	.7350	4	1.6305	2.9745	.6133
5	1.4693	3.9927	.6806	5	1.8424	3.5172	.5428
6	1.5869	4.6229	.6302	6	2.0820	3.9975	.4803
7	1.7138	5.2064	.5835	7	2.3526	4.4226	.4251
FOR INTEREST RATE 0.09 PER ANNUM				FOR INTEREST RATE 0.14 PER ANNUM			
1	1.0900	.9174	.9174	1	1.1400	.8772	.8772
2	1.1881	1.7591	.8417	2	1.2996	1.6467	.7695
3	1.2950	2.5313	.7722	3	1.4815	2.3216	.6750
4	1.4116	3.2397	.7084	4	1.6890	2.9137	.5921
5	1.5386	3.8897	.6499	5	1.9254	3.4331	.5194
6	1.6771	4.4859	.5963	6	2.1950	3.8887	.4556
7	1.8280	5.0330	.5470	7	2.5023	4.2883	.3996

Evaluation of Alternatives

YEAR	AMOUNT OF ONE DOLLAR AT COMPOUND INTEREST	PRESENT VALUE OF AN ANNUITY OF ONE DOLLAR	PRESENT VALUE OF ONE DOLLAR	YEAR	AMOUNT OF ONE DOLLAR AT COMPOUND INTEREST	PRESENT VALUE OF AN ANNUITY OF ONE DOLLAR	PRESENT VALUE OF ONE DOLLAR
FOR INTEREST RATE 0.15 PER ANNUM				FOR INTEREST RATE 0.2 PER ANNUM			
1	1.1500	.8696	.8696	1	1.2000	.8333	.8333
2	1.3225	1.6257	.7561	2	1.4400	1.5278	.6944
3	1.5209	2.2832	.6575	3	1.7280	2.1065	.5787
4	1.7490	2.8550	.5718	4	2.0736	2.5887	.4823
5	2.0114	3.3522	.4972	5	2.4883	2.9906	.4019
6	2.3131	3.7845	.4323	6	2.9860	3.3255	.3349
7	2.6600	4.1604	.3759	7	3.5832	3.6046	.2791
FOR INTEREST RATE 0.16 PER ANNUM				FOR INTEREST RATE 0.21 PER ANNUM			
1	1.1600	.8621	.8621	1	1.2100	.8264	.8264
2	1.3456	1.6052	.7432	2	1.4641	1.5095	.6830
3	1.5609	2.2459	.6407	3	1.7716	2.0739	.5645
4	1.8106	2.7982	.5523	4	2.1436	2.5404	.4665
5	2.1003	3.2743	.4761	5	2.5937	2.9260	.3855
6	2.4364	3.6847	.4104	6	3.1384	3.2446	.3186
7	2.8262	4.0386	.3538	7	3.7975	3.5079	.2633
FOR INTEREST RATE 0.17 PER ANNUM				FOR INTEREST RATE 0.22 PER ANNUM			
1	1.1700	.8547	.8547	1	1.2200	.8197	.8197
2	1.3689	1.5852	.7305	2	1.4884	1.4915	.6719
3	1.6016	2.2096	.6244	3	1.8158	2.0422	.5507
4	1.8739	2.7432	.5337	4	2.2153	2.4936	.4514
5	2.1924	3.1993	.4561	5	2.7027	2.8636	.3700
6	2.5652	3.5892	.3898	6	3.2973	3.1669	.3033
7	3.0012	3.9224	.3332	7	4.0227	3.4155	.2486
FOR INTEREST RATE 0.18 PER ANNUM				FOR INTEREST RATE 0.23 PER ANNUM			
1	1.1800	.8475	.8475	1	1.2300	.8130	.8130
2	1.3924	1.5656	.7182	2	1.5129	1.4740	.6610
3	1.6430	2.1743	.6086	3	1.8609	2.0114	.5374
4	1.9388	2.6901	.5158	4	2.2889	2.4483	.4369
5	2.2878	3.1272	.4371	5	2.8153	2.8035	.3552
6	2.6996	3.4976	.3704	6	3.4628	3.0923	.2888
7	3.1855	3.8115	.3139	7	4.2593	3.3270	.2348
FOR INTEREST RATE 0.19 PER ANNUM				FOR INTEREST RATE 0.24 PER ANNUM			
1	1.1900	.8403	.8403	1	1.2400	.8065	.8065
2	1.4161	1.5465	.7062	2	1.5376	1.4568	.6504
3	1.6852	2.1399	.5934	3	1.9066	1.9813	.5245
4	2.0053	2.6386	.4987	4	2.3642	2.4043	.4230
5	2.3864	3.0576	.4190	5	2.9316	2.7454	.3411
6	2.8398	3.4098	.3521	6	3.6352	3.0205	.2751
7	3.3793	3.7057	.2959	7	4.5077	3.2423	.2218

The comparison of the net present value computations for proposals A and B show that B, which has an NPV of -$2229.12, is better than A, which has an NPV of -$2625.64 since B is the most positive (least negative). In the example as given, the reduced energy cost does not offset the increased acquisition cost in the tax and interest rate environment of the company in the example.

If the interest rate were lower, say 12%:

Proposal A:

NPV = -2655 + 3.6048 (9.39)

= -2621.15

Proposal B:

NPV = -2059 + 3.6048(-54.4)

= -2255.10

Proposal B is still the best choice.

If the drive were to be operated two shifts, 4000 hr per year, and the interest rate were 12%:

Proposal A:

NCF_{annual} = -(4000 × 0.1696758 + 132.75)(0.52) + 531(0.48)

= -421.96 + 254.88

= -167.08

Proposal B:

NCF_{annual} = -(4000 × 0.19089 + 102.95)(0.52) + 411.80(0.48)

= -450.59 + 197.66

= -252.93

Proposal A:

NPV = -2655 + 3.6048(-167.08)

= -3257.30

Proposal B:

NPV = -2059 + 3.6048(-252.93)

= -2970.76

Proposal B, with its lower acquisition cost, is still the economic choice even though it will use $423 more electric power in the 5-year period covered by the comparative analysis.

It can be seen that if the tax rate were lower, or if the cost of money were lower, or if the cost of energy were higher, or if the intended usage were greater, the outcome of the analysis might have been different.

The energy rate chosen was a typical value for the Middle Atlantic states for the summer of 1981. Table 11.3 shows other rates which were typical for other regions of the United States at that time.

The analyst has to use the tax rates that apply to a particular company, the interest rate the company is capable of obtaining, and the energy rate the company is charged. The typical figures used in this section simply serve as values with which to carry on the illustration in the absence of more specific information.

11.5.1 Net Present Value Acceptance Criteria: Opportunity Cost

In proposals A and B of Section 11.5, both net present value (NPV) calculations produced negative numbers. Whereas the NPV method provides a succinct evaluation technique which takes into account the

TABLE 11.3 Typical Electric Utility Rates in the United States by Area, Summer 1981

Location	Rate for general lighting and power, 10,000 kWh/month (cents/kWh)
Middle Atlantic	8.7
New England	7.8
South Atlantic	6.43
West North Central	6.29
East North Central	6.24
West South Central	5.98
East South Central	5.86
Mountain	5.69
Pacific	5.23

Source: Courtesy of Edison Electric Institute, Washington, D.C.

company's tax bracket and its cost of funds, it is ordinarily used by accountants to signal acceptance or rejection of a project under review. As stated in Section 11.4, a negative NPV ordinarily means that the project will not recover the cost of money in the particular tax and interest environment.

When two projects are compared, such as traction drive proposal A or traction drive proposal B, the more positive of the two NPVs represents the best return for a company in that tax bracket and interest environment.

Practitioners of the NPV method hate to see negative numbers. An intangible "opportunity cost" should be added equally to both NPV evaluations. This represents a quantification of the company's decision to have a traction drive in the first place. The numerical value of the opportunity cost does not matter as long as it is big enough to make both NPVs positive, and as long as it is the same number for each NPV, so that it does not disturb the balance of the calculations. The company accountant will understand what has been done, and will be happy to concur in a selection based on positive NPV answers.

11.6 INTEREST RATE TABLES

11.6.1 Compound Interest

The discounted cash flow analysis is simply recognition that money has time value. It is a realistic method of comparing competitive investments by reflecting all the costs associated with them to one lump sum at the present time even though some of these costs are not to be paid out until some time in the future.

The general formula that governs discounted cash flow analyses is:

$$P_V = \frac{1}{(1+i)^t} \tag{11.4}$$

where

P_V = present value of a one-dollar payment to be made at a future time t
t = time in the future when the payment must be made, years
i = interest rate per year, expressed as a decimal

P_V must then be multiplied by the size of the payment involved, since the formula is for the present value of one dollar to be made in the future. Table 11.2 is a summary of the solution of this equation for

Interest Rate Tables 273

interest rates from 5 to 24% for times from 1 to 7 years, for interest compounded annually.

To illustrate, what is the present value of a payment of $1000 to be made 5 years from now if the interest is 12% per year compounded annually?

$$P_V = \frac{\$1000}{(1 + 0.12)^5}$$

$$= \$567.40$$

This number may be found in Table 11.2 as the present value of one dollar in the 12% section of the table.

If the analysis involves several payments to be made at different times, they each have to be calculated separately and added up to find the total present value. If all the payments are to be equal, the transaction is called an annuity.

11.6.2 Annuities

The present value of an annuity of one dollar to be paid out at the end of each year for t years is

$$P_V = \frac{1 - (1 + i)^{-t}}{i} \tag{11.5}$$

where

P_V = present value of the annuity of one dollar to be paid out at the end of each year
t = number of years for which the annuity is to run
i = interest rate per year expressed as a decimal

Equation (11.5) allows just one calculation to be made to reflect the present value of any equal number of future payments, instead of the multiple calculations that would have been required in the illustration of Section 11.5. The payments have to be made at the end of an interest period and there have to be the same number of payments as interest periods. Table 11.2 has a summary of annuity present values for rates from 5 to 24% for from 1 to 7 years.

If the payments have to be made annually, but the interest is to be compounded quarterly or at some other period than the payments, an equivalent interest rate must be used.

11.6.3 Equivalent Interest Rate

The equivalent interest rate is easily determined:

$$i_e = i \times \frac{(1 + i/n)^n}{1 + i} \tag{11.6}$$

where

i_e = equivalent interest rate to use in equation (11.5), expressed as a decimal
i = annual interest rate, expressed as a decimal
n = number of interest compoundings per year

Of interest to engineers is the fact that the engineering constant e (the base of natural logarithms, 2.718281828) is derived from the numerator of equation (11.6). e is the limit of the expression

$$\left(1 + \frac{i}{n}\right)^n$$

as n grows very large for the case where $i = 1$.

REFERENCE

1. R. Charles Moyer, James R. McGuigan, and William J. Kretlow, *Contemporary Financial Management*, West Publishing Company, St. Paul, Minn., 1981, p. 200.

12

DRIVE SELECTION: EXAMPLES AND CASE HISTORIES

12.1 INTRODUCTION

In this chapter several applications have been analyzed in detail using information developed in earlier chapters. In Section 12.2, on the high-speed blanking press, the flywheel design is integrated with the problem statement, and a motor selection is made. The complete startup transient of the press is traced by stepwise integration, and drive load peaks are evaluated.

In Section 12.3, on the municipal waste compactor, the consequences of an unexpected jam are examined. In Section 12.4, on the hosiery knitting machine, the special consideration of a fire hazard is solved. In Section 12.5, on the 100-hp paint mixer, the effect of changing product viscosity during batch processing is examined. The traction drive selected is shown to be dramatically more productive and efficient than the eddy-current drive it replaces. The sizing of a heat exchanger is examined so that the heat rejection of a high-horsepower drive can be accommodated. In Section 12.6, on the cutoff machine, special characteristics of very high speed-up ratios are examined.

The case of a traction drive that is part of a solar energy converter for a remote settlement is discussed in Section 12.7. In Section 12.8 there is also a general treatment of many machine tool and industrial applications. In each succeeding illustration only the aspects not previously treated are examined.

12.2 TRACTION DRIVE FOR A HIGH-SPEED BLANKING PRESS [1]

Specifications:

    Maximum load capability, 179,200 lb (80 metric tons)
    Maximum capacity available, 0.1 in. before bottom dead center
    Press stroke, 1.0 in.
    Drive, direct from press flywheel

The press in question is shown in Fig. 12.1. It has a coiled-strip continuous feeder which is not part of the problem statement. The flywheel is behind the shield at the upper right. The crankshaft runs horizontally in the top of the frame. The traction drive is mounted above the press. Essentially, the press consists of a reciprocating slide driven by a connecting rod and a crankshaft with a large flywheel.

The punch, attached to the slide, enters a die mounted on the bed of the press, thus shearing the workpiece. The material to be sheared will not be more than 0.1 in. thick. The rated force output of the press can only be developed in the last 0.1 in. of its stroke, for that is when the mechanical advantage of the crank mechanism is most favorable. The angular travel of the flywheel during the last 0.1 in. of stroke can be calculated from the details specified above to be 37°, in accordance with Fig. 12.2.

The remaining angular travel is expended withdrawing the punch and speeding up the flywheel. After punch withdrawal, 286° of flywheel angular travel is available for stock feed before the punch reenters the workpiece. The punched blank is ejected at the bottom.

The press has a clutch that will disengage the slide from the crankshaft and hold it at the top of each stroke. The crankshaft will continue to turn freely until the clutch is reengaged, when it carries the punch down again. Automatic mechanisms may prevent the clutch from reengaging if the guard is not in place or the workpiece is not positioned properly.

The traction drive has to be designed on the basis that punching action will be continuous. All the energy withdrawn from the press in order to punch the workpiece during the 37° angular working travel has to be restored before the same point in the next cycle. The traction drive has 360° of angular travel to do this in because the drive will be working during the entire stroke, not just during the 323° recovery period.

    Step 1. Calculate the motor horsepower. Work done per stroke at maximum force rating:

Traction Drive for a High-Speed Blanking Press

FIG. 12.1 High-speed blanking press. (Courtesy of F.U. (London) Ltd., Leicester, England.)

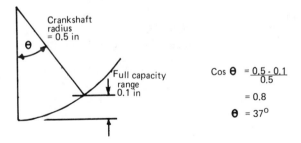

FIG. 12.2 Punch stroke geometry.

Work = force × distance

= 179,200 lb × 0.1 in.

= 17,920 in. lb

= 1493.33 ft lb

Maximum number of strokes per minute = 300

Work per minute = work/stroke × strokes/min

= 1493.33 × 300

= 448,000 ft lb min^{-1}

$$\text{Horsepower} = \frac{\text{work done per minute}}{33,000 \text{ ft lb min}^{-1} \text{ hp}^{-1}}$$

$$= \frac{448,000}{33,000}$$

$$= 13.57$$

We have not yet selected a traction drive. From the above we know the average power level at the punch point. The press has an efficiency associated with its mechanism and so does the traction drive. We make a tentative assessment of 75% for the product of these two factors.

$$\text{Motor horsepower required} = \frac{\text{press horsepower}}{\text{efficiency}}$$

$$= \frac{13.57}{0.75}$$

$$= 18.1$$

Step 2. Select the motor size. From Table 8.3, note that NEMA motors are available in nominal ratings of 15 and 20 hp but not 18 hp. Therefore, a 20-hp motor is tentatively selected.

Note that in Table 9.1, punch presses, listed under machine tools, have a service factor of 3. This implies that peak torque exceeds average running torque by a factor of 3:1. A brute-force approach would be to oversize the motor and the drive by a factor of 3. That would be grossly uneconomical and cannot be ranked as a solution. Instead, the characteristics of traction drives can be used to solve the problem elegantly with a 20-hp drive.

Step 3. Utilize available power.

Available power = motor horsepower × drive efficiency

Traction Drive for a High-Speed Blanking Press

$$= 20 \times 0.75$$
$$= 15 \text{ hp}$$

Step 4. Size flywheel. The flywheel must operate over the range 100 to 300 rpm. Energy storage in a rotating mass is a function of the square of its speed. The flywheel will have minimum energy available at its lowest speed, 100 rpm. The energy required of the flywheel is equal to the energy needed by the punch stroke minus the energy available from the motor during the punch stroke.

Energy of each punch stroke = 1493.33 ft lb

In this context energy, ft lb, is work, ft lb, since work is the useful form of mechanical energy. The energy available from motor during the punch interval is

$$E_{motor} = 15 \text{ hp} \times 33{,}000 \text{ ft lb min}^{-1} \text{ hp}^{-1} \times 0.01 \text{ min stroke}^{-1} \times \frac{37°}{360°}$$

$$= 508.75 \text{ ft lb}$$

Energy required from the flywheel = 1493.33 - 508.75

$$= 984.58 \text{ ft lb}$$

Good practice allows flywheels to slow down about 10% while yielding up their energy. Since flywheel energy is proportional, among other things, to speed squared, the delivered energy between full speed, 100 rpm, and 90% speed, 90 rpm, is

$$KE_{delivered} = KE_{total}(1.00^2 - 0.90^2)$$
$$984.58 = KE_{total}(1.00 - 0.81)$$
$$KE_{total} = 5182 \text{ ft lb}$$

The kinetic energy in a rotating mass is

$$KE = \tfrac{1}{2}Wk^2 \times \omega^2$$

where Wk^2 is the flywheel inertia from Table 4.2 and ω is the flywheel speed in rad sec^{-1}. For 100 rpm,

$$\omega = \frac{100 \times 2 \times \pi}{60}$$

$$= 10.47 \text{ rad sec}^{-1}$$

The required Wk^2 can be solved from the equation above.

$$Wk^2 = \frac{2 \times KE}{\omega^2}$$

$$= \frac{2 \times 5182 \text{ ft lb}}{10.47^2 \text{ rad}^2 \text{ sec}^{-2}}$$

$$= 94.5 \text{ ft lb sec}^2$$

In Table 4.2, Wk^2 is tabulated in terms of lb ft^2. To convert 94.5 ft lb sec^2 to lb ft^2, it must be multiplied by the acceleration of gravity, 32.17 ft sec^{-2}.

$$Wk^2 = 94.5 \text{ ft lb sec}^2 \times 32.17 \text{ ft sec}^{-2}$$

$$= 3040 \text{ lb ft}^2$$

From Table 4.2, column 2:

$$Wk^2 \text{ of a cylinder} = \frac{\pi r^4 t \rho}{2}$$

where

r = radius, ft
t = thickness, ft
ρ = density, lb ft^{-3}, from Table 4.1; note that the density of iron (or steel, which is mostly iron) is 486 lb ft^{-3}

After some trial, a flywheel of outside diameter 42 in. (3.5 ft), inside diameter 30 in. (2.5 ft), and a thickness of 7.5 in. (0.625 ft) yielded the following Wk^2 calculation:

$$Wk^2 = \frac{\pi \times [(3.5/2)^4 - (2.5/2)^4] \times 0.625 \times 486}{2}$$

$$= 3310 \text{ lb ft}^2$$

which is about 10% greater than the theoretical requirement, and therefore entirely satisfactory. The hub and shaft of the flywheel have been ignored. Because the inertia is sensitive to diameter to the fourth power, the small-diameter parts contribute very little to the overall inertia.

Step 5. Select drive speed. This is a belt-driven press. The output speed of the drive is established by the size of the drive pulley, since the flywheel is the load pulley and its size has been established in step 4. Brief mention was made in Section 11.5 of the validity of reducing drive cost by running its output as fast as possible within its rating. The linear reduction of life due to the increased number of stress repetitions in a given amount of time is offset by the increase of life, which is inversely proportional to the cube of the load torque, which is reduced.

This implies that the drive pulley should be small. The pulley has to meet the 20-hp power transmission requirements. The designers of this press chose a 10-in. pulley. In conjunction with the 42-in. flywheel, the drive output speed will be 42/10 faster than the flywheel. For the flywheel speed range of 100 to 300 rpm, the drive will operate at an output speed of 420 to 1260 rpm.

Step 6. Select motor characteristic. When the flywheel has slowed 10%, assuming that the drive has not changed ratio or slipped, the motor has also been forced to slow 10%.

Figure 8.6 shows that a standard NEMA Design B motor develops full-load torque at about 97% of synchronous speed. If forced to drop to 87.3% of synchronous speed (90% × 97% = 87.3%) it will develop 180% of normal full-load torque. Most NEMA motors have a 1.15 service factor built into their ratings. They cannot be called on to produce 1.8 times normal full-load torque more than one or two times per hour without experiencing severe heating.

Figure 8.9 shows that a NEMA design D characteristic should be selected. A NEMA design D motor produces full-rated torque at about 90% of synchronous speed. If forced to slow down to 81% of synchronous speed, it will produce about 130% of normal rated torque. The average speed during the punch cycle will be 95% speed. The flywheel is assumed to slow down linearly from 100 rpm to 90 rpm during the 37° punch operation, then to recover linearly to 100% speed in the remaining 323°. Since both slowdown and speedup operation segments are linear, the average speed should be halfway between the highest and the lowest points. If the Nema Design D motor is set up so that it produces full-load torque when the drive is operating at 95 rpm, the excursion will be -5% to minimum speed and +5% to maximum speed. From Fig. 8.9 it can be seen that the torque pulses will be about ±15%, within the service factor of the motor.

Step 7. Select motor speed. Candidate motor speeds are usually 1800, 1200, and 900 rpm. Since the drive output speed was established as 420 to 1260 rpm, the required drive ratio can be calculated. The inertia of the flywheel reflects back to the motor as the square of the ratio of flywheel speed to motor speed. The following tabulation can be constructed:

Motor speed (rpm)	Drive ratio required		Flywheel inertia, Wk^2, reflected to motor at:	
	420 rpm	1260 rpm	Lowest ratio	Highest ratio
1800	4.29	1.43	10.21	91.94
1200	2.86	0.952	22.99	206.88
900	2.14	0.714	40.86	367.78

Consult Fig. 8.12. Entering at 20 hp, note that an 1800-rpm motor can start at about 120 lb ft^2 load inertia, a 1200-rpm motor can start at about 300 lb ft^2 load inertia, and a 900-rpm motor can start at about 640 lb ft^2 load inertia.

Since the ability to start the load, even if the ratio were left in the maximum position, would not be a problem with any of the candidate motor speeds, motor speed selection should be based on the ratio capability of the candidate drives. Many drives are advertised in terms of the product of their highest speed divided by the lowest speed. A 6:1 drive which can go from one-third of input speed to two times input speed does not do you any good if you have an output requirement of 1/4.29 to 1/1.43. The ratio needed is only 3:1 but it is all in one direction, and the 6:1 drive just mentioned could not meet the lowest speed.

Candidate drives that were considered were the V20 and the V30 from the F.U. variable-speed-drive catalog. The V20 has the required speed ratio capability, 3:1, but it is balanced entirely below the input speed. To achieve the 1260-rpm-drive output speed selected in step 5, the V20 would have to be operated by a 1200-rpm motor and would have to step up speed slightly at the highest ratio. If the V20 were to be a serious candidate, the pulley size would have to be adjusted to get the speed within range. As will be shown in step 9, if the pulley is to be resized, it might be better to make it small enough so that the drive could operate at 1800 rpm instead of at 1200 rpm, where it has a slightly lower power capacity.

The V30 drive has a speed ratio of 5.29 to 0.882. With this drive we could have selected the 1800-rpm motor or the 1200-rpm motor without changing the pulley size. We would select the 1800-rpm motor because four pole is a more common and lower-cost construction than six pole, and because the input torque for 20 hp would be less for the higher speed.

Motor selection has to be made based on both motor characteristics and available drive characteristics. If the V30 were selected based on the parameters examined up to this step, the 1800-rpm motor would be chosen.

Step 8. Analyze the starting transient. Assume that the drive will normally be started in its lowest working ratio, 4.06:1, so that motor speed will be 1620 rpm when the flywheel is at 95 rpm. Load inertia is 3310 lb ft^2. Reflect this back to the drive motor at the square of the overall ratio.

$$Wk^2_{\text{load reflected}} = \frac{3310}{(4.06 \times 4.2)^2}$$

$$= 11.38 \text{ lb ft}^2$$

Either redraw Fig. 8.9 on the basis that 100% torque equals 64.84 lb ft and 100% speed equals 1800 rpm, or read the figure as it stands and multiply the percentage figures noted. As a check, note that the NEMA design D motor develops 100% torque at 90% speed. Its horsepower would be

$$\text{hp} = \frac{64.84 \times 1800 \times 0.9}{5252}$$

$$= 20$$

Motor inertia may be estimated from Fig. 8.11. Enter at 20 hp and read $I_{\text{motor}} = 2.4$ lb ft^2 from the four-pole curve.

The procedure for analyzing the transient is as follows:

a. Speed N = zero. Read Fig. 8.9: T = 182 lb ft.

b. Acceleration $= \dfrac{T \times \text{efficiency}}{(I_{\text{load}} + I_{\text{motor}})/32.17}$

$= \dfrac{182 \times 0.75}{(11.38 + 2.4)/32.17}$

$= 318 \text{ rad sec}^{-1}$

$= 3041 \text{ rpm sec}^{-1}$

For stepwise integration using steps of 0.1 sec, the speed change is $0.1 \times 3041 = 304$ rpm sec^{-1}.

c. New speed = old speed + speed change

= zero + 304 rpm

= 304 rpm

Repeat the steps above for time t = 0.1 sec:

1a. Read Fig. 8.9 for 304 rpm: T = 170 lb ft.

1b. $a = \dfrac{170 \times 0.75}{13.78/32.17}$

$= 298 \text{ rad sec}^{-1}$

$= 2849 \text{ rpm sec}^{-1}$

$= 284.9 \text{ rpm in } 0.1 \text{ sec}$

1c. $N = 304 + 285 \text{ rpm}$

$= 589 \text{ rpm}$

The process is repeated until the speed equals the full-load speed of the motor. At that point the starting transient is considered complete. The steps would look as follows:

2a. T = 157 lb ft
2b. ΔN = 262 rpm/0.1 sec
2c. N = 851 rpm
2d. t = 0.2 sec

3a. T = 139 lb ft
3b. ΔN = 232 rpm/0.1 sec
3c. N = 1082 rpm
3d. t = 0.3 sec

4a. T = 122 lb ft
4b. ΔN = 203 rpm/0.1 sec
4c. N = 1286 rpm
4d. t = 0.4 sec

5a. T = 107 lb ft
5b. ΔN = 178 rpm/0.1 sec
5c. N = 1464 rpm
5d. t = 0.5 sec

6a. T = 89 lb ft
6b. ΔN = 148 rpm/0.1 sec
6c. N = 1612 rpm
6d. t = 0.6 sec

7a. T = 63 lb ft
7b. ΔN = 105 rpm/0.1 sec
7c. N = 1717 rpm
7d. t = 0.7 sec

The starting transient analysis showed that the 20-hp motor would get the flywheel up to 95 rpm (1620 rpm of the drive motor) in 0.6 to 0.7 sec.

Traction Drive for a High-Speed Blanking Press

In the groups of numbers above the torque for the next group is read from Fig. 8.9 for the speed of the last group, and is considered constant over the next 0.1 sec, during which a substantial speed change takes place. The error created by the use of this assumption is not very great if reasonable-size steps are taken. Using a small computer, the problem above was rerun using 10 times as many steps, each 0.01 sec. The time to reach 1620 rpm was shown to be 0.755 sec. This is an excellent starting time. The motor should have no startup heating problems.

The torque experienced by the drive input is slightly less than the torque of the motor as read from Fig. 8.9. This is because the motor is accelerating itself as well as the load inertia reflected to it. Some of the motor torque is expended on its own inertia. This torque fraction is not seen by the drive. The peak torque seen by the drive is

$$T_{\text{drive peak}} = \frac{T_{\text{motor locked rotor}} \times I_{\text{reflected load}}}{I_{\text{reflected load}} + I_{\text{motor}}}$$

$$= \frac{182 \text{ lb ft} \times 11.38 \text{ lb ft}^2}{(11.38 + 2.4) \text{ lb ft}^2}$$

$$= 150 \text{ lb ft}$$

The drive manufacturer has to accept this peak torque, seen each time the press is started, as part of the drive load spectrum. It must be agreed that the drive will not slip grossly, nor will it be overstressed in any way that would diminish its life.

If the problem is restated in terms of reflected inertia being increased by a factor of 9, from 11.38 to 102.42, we can see what would happen if the press were inadvertently started up in minimum drive ratio, so that the flywheel would come up to 300 rpm when the motor comes up to 1620 rpm. The computer reports that start time to 1620 rpm will be 5.75 sec. This is a very slow start and motor heating will be significant. 102 lb ft^2 is just above the limiting connected inertia recommended by NEMA for a single start of a 20-hp motor when at operating temperature, or for two starts, with the motor coasting to a stop in between, if the motor was cold initially. Figure 8.12 shows connected inertia limits recommended by Westinghouse for a range of motor speeds and horsepowers.

When the drive is started in low ratio, the torque impressed on the drive by the motor will be greater than when it is started in high ratio. This is because the reflected load inertia is so much higher that, proportionally, less of the motor torque is expended accelerating the motor and more is expended accelerating the load. The peak torque seen by the drive in this case is

$$T_{\text{drive peak}} = \frac{T_{\text{motor locked rotor}} \times I_{\text{reflected load}}}{I_{\text{reflected load}} + I_{\text{motor}}}$$

$$= \frac{182 \text{ lb ft} \times 102.42 \text{ lb ft}^2}{(102.42 + 2.4) \text{ lb ft}^2}$$

$$= 178 \text{ lb ft}$$

Since, according to one well-known engineering sage, if something can go wrong, it will, it would be best not to rely on the press operator always reducing the speed to the minimum just before shutting the press down so that it will be in maximum ratio for startup. Many traction drives cannot change ratio when at rest, so that the speed ratio they were last run at is the ratio they have to start up with. The drive manufacturer should be expected to take the higher torque peak associated with starting the drive in minimum ratio. If it is regarded as a desirable feature, the drive manufacturer would be prepared to furnish a servo-operated speed-adjust mechanism which would crank the drive down to lowest output speed during the coast-down period once the power had been cut.

Step 9. Calculate the maximum speed-change servo rate. The 20-hp motor generates 20 × 550 or 11,000 ft lb per second. The flywheel has a Wk^2 of 3310 lb ft^2 or 102.42 lb ft sec^2. The kinetic energy stored in the flywheel is

$$KE = \tfrac{1}{2} I \omega^2$$

where I is Wk^2/g in lb ft sec^2 and ω is speed in rad sec^{-1}. So

$$KE = 0.005483 I N^2$$

where N is speed in rpm. For this flywheel

$$KE = 0.564 N^2$$

At 300 rpm, KE = 50,760 ft lb; at 301 rpm, KE = 51,099 ft lb. The difference, 339 ft lb, is the energy required to increase flywheel speed 1 rpm.

The motor can deliver 11,000 ft lb sec^{-1} × 0.75 = 8250 ft lb sec^{-1} when the drive system efficiency is taken into account. It follows that this motor can accelerate the flywheel at

$$\frac{8250}{339} = 24 \text{ rpm sec}^{-1}$$

Similar calculations, done for 200 and 100 rpm, where the flywheel contains less energy, show that the maximum acceleration rates can be increased to 36 rpm sec^{-1} at 200 rpm and 72 rpm sec^{-1} at 100 rpm. The speed control servo has to be designed to respect the limiting speed change rates.

When speed is reduced sharply, the drive, increasing in ratio, could tend to overspeed the motor if the flywheel tends to continue rotating at the old speed. The motor can remove energy from the rotating system by its friction and windage loss and by pumping some energy back into the power line as an induction generator. Check that the drive would not be damaged by brief reverse loading. Then make sure that the servo speed change is introduced gradually enough to avoid the problem entirely.

Step 10. Check with the factory. The traction drive selected for this press by the techniques of steps 1 to 8 will have been very conservatively rated, and will provide satisfactory service.

The factory, in publishing its nominal drive ratings, will have established conservative limits to protect those who do their own sizing. Sometimes it is possible that, after consideration of all factors, a unit one frame size smaller would also be able to give satisfactory service. The illustration given in Section 11.5 was actually one such case where the factory was able to propose a 22% cost saving by offering a smaller-frame-size unit which ran at higher speed but which did the task equally well.

Figure 12.3 is a chart published by F.U. (London) Ltd., which offers some insight into other factors that relate to drive selection. This chart relates:

1. Number of operating hours per day
2. Loading conditions during starting
3. Number of starts per hour
4. Ratio of peak torque to average torque

For purposes of illustration, assume that the drive for the high-speed blanking press described in this section is expected to encounter the following operating conditions:

1. Number of operating hours per day, 8
2. Peak torque seen during starting (from step 8), 234%
3. Number of starts per hour, less than 1
4. Ratio of peak torque to average torque (step 6), 1.15

Enter Fig. 12.3 at a point along the horizontal axis which corresponds to the number of hours per day; in this case, 8 hr per day would be used.

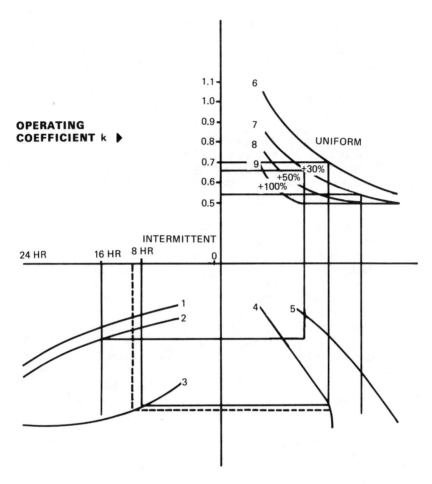

FIG. 12.3 Nomograph for operating coefficient k. (Courtesy of F.U. (London) Ltd., Leicester, England.)

Draw a vertical line downward to intercept curves 1, 2, or 3.

Curve 1 applies if the traction drive is started under no-load conditions, with the driven machine remaining at rest.
Curve 2 applies if the traction drive is started under load but the peak torque during starting does not exceed 120% of normal running torque.
Curve 3 applies if the traction drive is started under load but the peak torque during starting does not exceed 230% of normal running torque.

Traction Drive for a High-Speed Blanking Press

The problem statement we are considering requires us to intercept curve 3.

Next draw a horizontal line from the intercept with curve 1, 2, or 3, to intercept curve 4 or 5.

> Curve 4 applies if the number of starts is less than 1 per hour.
> Curve 5 applies if the number of starts is 10 or more per hour.

The analyst may interpolate between curves 4 and 5 if the number of starts per hour is an intermediate value.

Next draw a vertical line up from the intersection with curve 4 or 5 to intercept curve 6, 7, 8, or 9.

> Curve 6 corresponds to uniform loading with virtually no torque peaks above the average running torque.
> Curve 7 is used if surges or torque peaks occur which are on the order of 20 to 30% above the average running torque.
> Curve 8 is used if surges or torque peaks occur which exceed average running torque by 50%.
> Curve 9 is used where there are surges of 100% above average running torque, where rotation of the drive is frequently reversed, or where there is a high frequency of starts coupled with a high inertia.

When a line is drawn starting from 8 hr per day, to curve 3, then to curve 4, then to curve 7, and across to the vertical axis, an operating coefficient k may be determined. In this case k will be found to be 0.6.

The factory has the following output horsepower ratings for candidate drives that might be used for this application:

Drive number	Hp at 1400-rpm input		Hp at 1800-rpm input	
	500-rpm output	1500-rpm output	600-rpm output	1800-rpm output
V20	21	25	23	27
V30	30	36		

Since the drive is assumed to be 90% efficient, the corresponding input horsepowers are:

V20	23	27	25.5	30
V30	33.3	40		

290   *Drive Selection*

In step 2 we had selected a 20-hp motor. Because the operating coefficient k was determined to be 0.6, the drive input horsepower rating would have to be 20/0.6 or 33.3 hp.

The factory might suggest:

1. Use a star-delta starting system for the motor. This would reduce motor torque during starting, and would enable the operating coefficient k to be based on curve 2 of Fig. 12.3. When the operating coefficient is recalculated from the nomograph of Fig. 12.3, k is now found to be 0.77 and the required drive input horsepower rating need only be 20/0.77 = 26 hp.
2. Reduce the drive output pulley from 10 in. to 7 in. in diameter. This changes the required drive speed from 1260 rpm to 1800 rpm, thus validating the increased rating of the drive.
3. Use a V20 drive. The small rating difference at 600 rpm will not matter.

In 1980 dollars a V20 drive is about $4270 less expensive than a V30 drive. However, it would require the star-delta starting system, which would be about $1076 more expensive than a simple across-the-line contactor. On balance a substantial saving would have been made.

The factory, or its local representative, should always be contacted once some rough sizing has been done and the information required by the drive specification sheet (Chapter 10) has been assembled. Some of the data they may have, which may not have been put in the catalog, may save you money, or conversely, may protect you against a loading condition you may have overlooked.

The curves in Fig. 12.3 illustrate general relationships between the application variables. These data are for F.U. roller disc drives and may not necessarily apply to other makes of traction drives.

12.3 DRIVE FOR A MUNICIPAL WASTE COMPACTOR [2]

Specifications:

    Motor, 15 hp, totally enclosed fan-cooled, three phase, 60 Hz, 1800 rpm
    Speed reducer, in-line 35:1 ratio
    Offset chain reducer in final drive, 60/11
    Speed regulation, 1% at 1:1 ratio, 3.5% at 4.5:1 ratio
    Speed setting deviation, 0.02% at full load
    Remote speed adjustment (dc stepper motor or pneumatic servo)

The screws which these drives operate are designed to deliver shredded municipal waste (i.e., garbage) at rates ranging from 4000 to 16,500 lb per hour. Product density may run 10 to 25 lb ft^{-3}. The theoretical turndown ratio is approximately 10.3:1. Maximum torque at the screw shaft is calculated to be 126,000 lb in. However, at maximum output rate, the material density is generally at the low end of the range, which makes a constant-horsepower output desirable.

Other factors involved in the selection of these drives: the environment can be extremely dirty, which technically eliminates any speed-adjustment arrangement involving belts or sheaves; minimum drift from set operating speeds is desirable; maintenance in municipal operations may not be of general industrial quality, so extreme ruggedness of components is a factor; and finally, the probability of interference between screw and trough by oversize materials, causing stalling of the drive, is high.

In regard to this last point, simple overload protection is not considered adequate. It takes time to locate the point of interference, and even more time to remove the interfering object—perhaps as much as several hours. Consequently, electronic circuitry that senses clutch slippage and reverses the drive momentarily has been installed, with the expectation that such immediate reversal will permit the interfering item to drop down or favorably reposition itself with only a brief interruption in the continuity of the feed. In the operation at Albany [3], an unusually large number of oversize tramp materials, including large and long pieces of steel, have been discharged successfully and without interrupting the operation.

A diagram of the drive is shown in Fig. 12.4. The drives selected were Koppers series R size 16. These are roller drives that met the requirement for extreme ruggedness. The manufacturers rating for the R16 roller drive is:

Torque, 250 lb ft at 530 rpm, tapering to 57 lb ft at 277 rpm
Horsepower, 25 from 530 to 2150 rpm
Speed range, 277 to 2700 rpm
Set-point drift at 1800 rpm input, 0.032
Efficiency, 90% at 1:1 ratio, 84% at extreme ratio

The reducer is a Sumitomo Cyclodrive type H, frame size 59A. The slip clutch is a Kopp type UK4 with a slip indicator. The clutch setting is 990 lb in. The drive uses a NEMA design B motor characteristic. From Table 8.11 and Fig. 8.6, note that the peak torque ever achieved by a 25-hp NEMA design B motor is 200% of rated torque at about 75% of synchronous speed.

From equation (4.2) the rated torque of the motor is

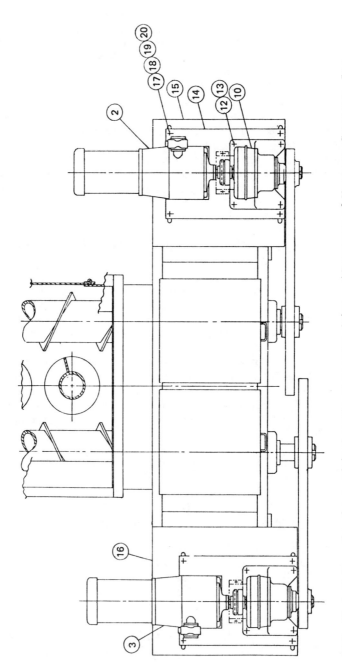

FIG. 12.4 Schematic of a municipal waste compactor drive. (Courtesy of Koppers Company, Inc., Sprout Waldron Div., Muncy, Pa.)

Drive for a Municipal Waste Compactor

$$T = \frac{5252 \times 15 \text{ hp}}{1740 \text{ rpm}}$$

$$= 45.28 \text{ lb ft}$$

Peak torque, from Fig. 8.6 and Table 8.11 (breakdown torque), is

$$T_{max} = 200\% \times 45.28$$

$$= 90.56 \text{ lb ft}$$

The clutch setting of 990 lb in. corresponds to 1.5 times the calculated screw torque as seen through the 35:1 reduction of the Cyclodrive multiplied by the 60/11 chain reduction.

It was noted in Section 9.5 that stopping could be the dominant factor in some traction drive applications. In this case, when a piece of tramp metal is ingested, the system may stop rather quickly. The clutch setting of 990 lb in reflects back to the motor as

$$T_{motor} = \frac{T_{clutch}}{\text{drive ratio} \times \text{efficiency}}$$

$$= \frac{990/12}{\text{ratio} \times 0.84} \text{ lb ft}$$

$$= \frac{98.2}{\text{ratio}}$$

From this and the motor breakdown torque calculated above, it can be seen that if the drive is in a ratio of 1.085 or less, the clutch can stall the motor, but if it is in a higher ratio, it cannot.

Consider a jam while the drive is in maximum speed mode. The drive ratio is 0.6667. At the time the jam is encountered, the torque reflected back to the motor from the slipping of the clutch is

$$T_{motor} = \frac{990 \text{ lb in.}}{12 \text{ in. ft}^{-1} \times 0.6667 \times 0.84}$$

$$= 147.3 \text{ lb ft}$$

From Fig. 8.6 it can be seen that the value of motor torque varies from approximately 140% (the pull-up torque) to 200% (the breakdown torque) for all speeds below about 95% speed. For convenience of illustration only, it may be assumed that the average motor torque during this transient from full speed to full stop is 160% of rated torque.

Deceleration is sometimes used as a word to describe the negative sense of acceleration. This book will not use that term, but will use acceleration and assign a negative value if the speed is being reduced.

$$T_{acceleration} = T_{motor} - T_{load}$$
$$= 1.6 \times 45.28 - 147.3$$
$$= -74.87 \text{ lb ft}$$

From Fig. 8.11 note that a 15-hp 1800-rpm motor has an inertia of 1.8 lb ft^2 approximately.

The acceleration of the motor is, from equation (4.6),

$$\alpha = \frac{torque}{inertia}$$
$$= \frac{-74.87 \text{ lb ft}}{1.8 \text{ lb ft}^2 / 32.17 \text{ ft sec}^{-2}}$$
$$= -1338 \text{ rad sec}^{-2}$$
$$= -12,778 \text{ rpm sec}^{-2}$$

This means that the system will come to a screeching halt within 0.136 sec if motor speed at the time the jam was encountered was 1740 rpm.

The clutch will slip during this interval. The motor will have turned:

$$\theta = \omega_0 t + \tfrac{1}{2}\alpha t^2$$
$$= 182.21 \times 0.136 + \tfrac{1}{2}(-1338) \times 136^2$$
$$= 12.37 \text{ rad}$$
$$= 1.97 \text{ rev}$$

where

ω_0 = initial motor speed, rad sec^{-1}
$= \dfrac{1740 \times 2 \times \pi}{60}$
t = stopping time, sec
α = acceleration, rad sec^{-2}

The clutch slip is

$$\text{Slip} = \frac{1.97 \text{ rev of motor to stop}}{0.667 \text{ drive ratio}} = 2.95 \text{ rev}$$

The work dissipated in the clutch is calculated in accordance with Section 4.2.2:

Work = 2 × π × 2.95 × 990

= 18375 in. lb

= 1531 ft lb

In accordance with Section 4.2.4,

1531 ft lb = 1.968 Btu

The clutch is 7.8 in. in diameter and 3 in. thick. A simple volume calculation yields a volume of 143 in.^{3} or 0.083 ft^3. From the fourth line of Table 4.1, we know that iron has a density of 486 lb ft^{-3} and a specific heat of 0.11 Btu lb^{-1} °F^{-1}. Thus the clutch weighs about 40.6 lb.

By the use of this information and equation (4.5), the temperature rise in the clutch would be

$$\Delta t = \frac{1.968 \text{ Btu}}{0.11 \times 40.6}$$

= 0.44 °F

The clutch does not mind the sudden stop. The slip value has been set based on the structural characteristics of the compactor screws and the reducer. The traction drive has to take the occasional peak torque reflected back to it when the compactor strikes a hard object. It does this very well.

The temperature rise calculated above occurred in 0.136 sec. This was an average temperature rise assuming that the heat diffuses into all the clutch metal. Such diffusion takes finite time. While it is happening, the clutch facings may be getting quite hot. This text does not go into unsteady-state heat transfer. That is left to the clutch manufacturer and more learned texts. It will suffice to say that the clutch facing might, momentarily, experience a temperature rise 10 times greater than that of the mean metal temperature rise.

If the drive were operating in a speed-reduction mode which allowed sufficient torque to slip the clutch continuously, the clutch might be destroyed in a short time if a jam were encountered. The fast response electronic circuit detects this slippage and reverses the drive to prevent subsequent clutch or motor damage.

The drive must be of the reversing type because each compactor has both a left-hand and a right-hand installation, each of which must turn in a different direction, and because the drives must be periodically reversed to clear the inevitable jams.

12.4 HOSIERY KNITTING MACHINE [4]

A factory with a large number of knitting machines desired to upgrade them from dc motor drive. The machines have approximately 1-hp drive requirements at maximum speed. The machines are programmed to operate at slow, medium, or high speed on demand at various points in the product cycle.

The mill environment in which the machines were used produced continuous oil and lint which totally covered the exterior of the drives with a 1-in. blanket, sometimes in as short a period as 24 hr. The dc drive motors used previously, even when very conservatively sized, had experienced considerable problems from overheating. The mill had had a history of numerous small fires started from overheated drive motors.

The knitting machines produce high-torque spikes as they shift from one function to another. Whatever solution was proposed had to accommodate these spikes without distress to the drive, its motor, or the control.

Figure 12.5 shows a row of machines equipped with the traction drive solution, which consisted of a Contraves model H25A 1.2-hp drive belt driven from a side-mounted totally enclosed fan-cooled (TEFC) motor similar to that shown in Fig. 8.2. Figure 12.6 illustrates the arrangement. The belt drive is a 1:1 arrangement. The belt is tensioned by shifting the motor laterally on its slotted foot mount. The output of the drive is reduced in speed by a ratio of 6:1 in the helical gear reducer fitted to the output shaft.

In the Contraves drive the speed ratio is controlled by the position of a central shaft which steers the control rollers. The drive motor is side mounted so that it does not block access to the control shaft which emerges from the input end.

The ratio of the speed variator was changed automatically on command from the knitting machine by a 200-pole Slo Syn stepping motor as illustrated in Fig. 12.7. A mechanical cam sequence timer was used to define angular positions which correspond to slow, medium, and fast machine speeds.

In 8 years no fires and no failures due to overheating have occurred. The drives have proved able to run for long periods of time at speeds as low as 10% of maximum running speed without encountering the problems experienced by the dc drives.

Experience has shown that the traction drive installation was easily serviced by factory maintenance mechanics. They could change lubricant, replace bearings, and perform other maintenance tasks. Traction drives exhibit a benign failure mode. They tend to warn of wear by becoming noisy before they fail catastrophically, and they can be serviced promptly by the personnel who maintain the knitting

FIG. 12.5 Hosiery knitting machines with traction drives. (Courtesy of TEK Products Corp., High Point, N.C.)

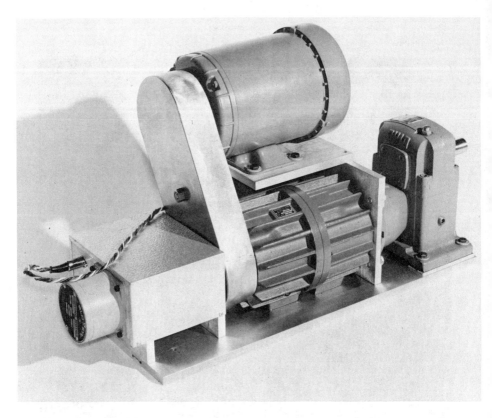

FIG. 12.6 Hosiery knitting machine drive assembly. (Courtesy of TEK Products Corp., High Point, N.C.)

machines. There had been concern that had an electronically controlled drive been selected, the necessary maintenance skills would not have been available.

12.5 100-HORSEPOWER PAINT MIXER [5]

Most paints are white or light in color. Large batches of a white stock base are produced, and then portions are tinted with various pigments to produce smaller batches of the colors required. In the production of the base stock, the lower-viscosity liquids are placed in a tank and agitated while the various components are added. As the solids build up in concentration, the viscosity of the batch builds

FIG. 12.7 Stepper motor servo for knitting machine traction drive. (Courtesy of TEK Products Corp., High Point, N.C.)

up from approximately 1 cP for the liquid vehicle to several thousand centipose for the finished paint.

Viscosity has been defined, technically, in Section 4.3. A characteristic of viscous drag is that the force to shear the fluid is directly proportional to velocity. If a constant-velocity mixing process takes place during which the viscosity of the mix increases, the retarding force on the mixer would increase proportionately.

Once the retarding force caused the mixer to reach its power limit, or its torque limit, mixing could be continued in the presence of further increases in viscosity only by allowing the mixer to slow down. In the case at hand, a 4000-gal paint mixing vat required a 100-hp mixer.

The original, technically successful, installation utilized an eddy-current drive. The motor ran at constant speed. The drive could provide torque up to the motor torque rating, which for an 875-rpm motor was 600 lb ft. The mixer was such that, to avoid overloading the motor, the eddy-current drive was controlled to slip so that the output shaft slowed down as the viscosity increased above a certain value, so that the motor torque demand did not exceed 600 lb ft.

One economic problem with eddy-current drives is that output torque always equals input torque, but output speed is always less than input speed. Therefore, output horsepower is always less than input horsepower. The difference is dissipated as heat into the cooling air.

When a 100-hp eddy-current drive is fully loaded and the output is slowed to half speed, the energy wasted will be exactly 50 hp. At the electric rate of Table 11.3 for a typical Middle Atlantic state, such wasted power amounts to $3.82 per hour if the motor generating it were 85% efficient. If the mixer worked one shift 5 days a week at this operating condition (2000 hours per year), the wasted power will have cost $7635.

A traction drive is the ideal solution for this type of application because a drive may be selected which can be run at constant horsepower. Then, as the viscosity of the batch increased, the necessity for increased torque could be met by changing the drive ratio so that the product of output torque and output speed remained a constant. The output torque would increase, above the motor torque, as the output speed is reduced. This is something an eddy-current drive cannot do.

Table 12.1 charts the comparison between an eddy current drive and a constant-horsepower traction drive for a scenario in which the viscosity increases by a factor of 9 from the value that fully loads the motor in both cases. Note that when the viscosity has increased to 900% of the limiting value, the traction drive is mixing three times faster than the eddy-current drive, yet both drives are fully loading their respective motors.

The mean advantage of using a traction drive for the paint mixer application was that the paint manufacturer was able to mix 8000-gal batches using the traction drive with the same 100-hp motor, in the same time, and with the same labor and energy cost, as had previously only been adequate to mix 4000-gal batches.

Figure 12.8 shows the solution as installed. The 100-hp motor is foot mounted at the right. It is coupled to the multiple-disc traction drive in the center. The mixing head is at the left. The entire assembly is mounted above the 8000-gal vat. The drive shaft passes through the cover of the vat.

TABLE 12.1 Comparison of Eddy-Current Drive and Traction Drive in a Mixer Application in Which Viscosity Increases During Mixing[a]

Viscosity	Eddy-current drive				Traction drive			
	Torque	Speed	Power input	Power output	Torque	Speed	Power input	Power output
100	100	100	100	100	100	100	100	100
200	100	50	100	50	141	71	100	100
300	100	33	100	33	173	58	100	100
400	100	25	100	25	200	50	100	100
500	100	20	100	20	224	45	100	100
600	100	17	100	17	245	41	100	100
700	100	14	100	14	265	38	100	100
800	100	13	100	13	283	35	100	100
900	100	11	100	11	300	33	100	100

[a]All values are percentages.
　This is a normalized table which does not recompute efficiency of the traction drive at each point. It assumes that at the 100% viscosity operating point both drives have equal efficiency and deliver equal power to the load.

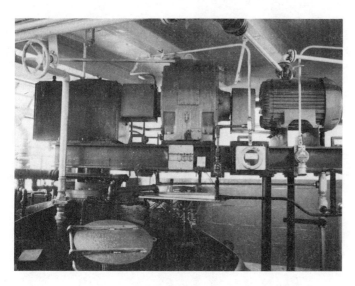

FIG. 12.8 100-Hp paint mixer installation. (Courtesy of Sumitomo Machinery Corporation of America, Teterboro, N.J.)

FIG. 12.9 Input end, 100-hp paint mixer traction drive. (Courtesy of Sumitomo Machinery Corporation of America, Teterboro, N.J.)

Figure 12.9 is a view of the motor and the input side of the drive. Note the external oil cooler on the right. The efficiency of the large multiple-disc traction drives ranges from 75 to 87% depending on the particular unit and operating ratio. If the efficiency were 85%, then 15% of 100 hp would be lost as heat to the traction lubricant. From Section 4.2.3, where 1 hp was defined as 33,000 ft lb min^{-1}, and from Section 4.2.4, where 1 Btu was defined as 778.26 ft lb, it can be deduced that 1 hp of heat rejection would be equal to

$$Q = \frac{33,000 \text{ ft lb min}^{-1} \text{ hp}^{-1}}{778.26 \text{ ft lb Btu}^{-1}}$$

$$= 42.4 \text{ Btu min}^{-1}$$

For 15 hp of heat rejection,

$$Q = 15 \times 42.4$$

$$= 636 \text{ Btu min}^{-1} \text{ or } 38,160 \text{ Btu hr}^{-1}$$

If a water-cooled heat exchanger were used, and the water were allowed to heat up 25°F (13.9°C), then

Cutoff Machine 303

$$W_{water} = \frac{Q}{c \times \Delta t}$$

$$= \frac{636}{1 \times 25}$$

$$= 25.44 \text{ lb min}^{-1}$$

$$= 3.05 \text{ gal min}^{-1}$$
(1 gal of water weighs 8.345 lb)

where

W = weight of water, lb min^{-1}
Q = heat rejection, Btu min^{-1}
c = specific heat of water, Btu lb^{-1} °F^{-1} (from Table 4.1)
Δt = temperature rise, °F

There is no question about the ability to air cool such a large drive. This text does not go into the mathematics of heat transfer. When free-convection calculations are made, it will generally be seen that for surfaces of a scale of 2 or 3 ft in free air the heat transfer coefficient will be from 1 to 2 Btu hr^{-1} ft^{-2} °F^{-1}. A brief examination of the surface area of the drive compared to the heat rejection requirement of 38,160 Btu hr^{-1} will quickly allow the engineer to conclude that forced cooling will be necessary to avoid excessive internal temperatures.

The high power level of this application highlights one area in which traction drives are superior to presently available competitive systems. The smaller sizes face competition from electronically controlled motors. The electronic controllers use transistors or similar devices which are not economically available for large output power levels. The energy savings over eddy-current or similar slip control drives are dramatic at the 100- to 200-hp level. They are much less dramatic when the drive is 1 or 10 hp instead of 100 or 200 hp.

12.6 CUTOFF MACHINE [6]

Traction drives may be operated as step-up drives for special applications. Figure 12.10 shows a traction-drive-operated cutoff machine powered by a 25-kW (33 hp) 1500-rpm motor. The output speed range is 2500 to 8000 rpm.

This is an application of the Kopp type K roller variator (Fig. 12.11) in which the input is shaft K and the output is shaft A. In the Kopp variators the shaft at the flanged end is usually the input shaft, except for high-speed application such as those described in

FIG. 12.10 8000-rpm cutoff machine. (Courtesy of Kopp-Variatoren AG, Mürten, Switzerland.)

Cutoff Machine

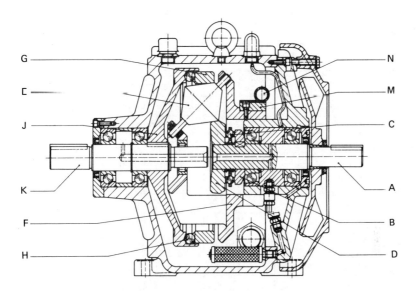

FIG. 12.11 Schematic of the Kopp series K traction drive. (Courtesy of Kopp-Variatoren AG, Mürten, Switzerland.)

this section. Note that the input and output shafts turn in opposite directions. Units to be run at very high speed require special high-precision parts, and are identified by the series HS. The traction drive shown in Fig. 12.10 is a KHS 16. Such units require external oil coolers because they are rated to handle substantial power. The fan shown in cross section in Fig. 12.11 is not required when an oil cooler is used. The fan has been omitted from the drive of Fig. 12.10.

Figure 12.12 is a performance chart for the type KHS traction drives operated in the step-up mode. Note the substantial power output ratings. Drives of this type exhibit two characteristics that electronically speed controlled motors cannot achieve: high power output and substantial speed increase.

This cutoff machine is used to trim concrete blocks and various ferrous materials. Output speed is selected in accordance with the material being cut. Output speed, once set, holds constant over long periods of time. The drive is rated to maintain speed within 0.022 to 0.004% of initial set point, depending on load and ratio.

This cutoff machine is subjected to 6 to 12 starts per day. During each start there is a momentary load peak estimated to be three times the normal running torque. It has been in service 14 years, and has yet to be overhauled.

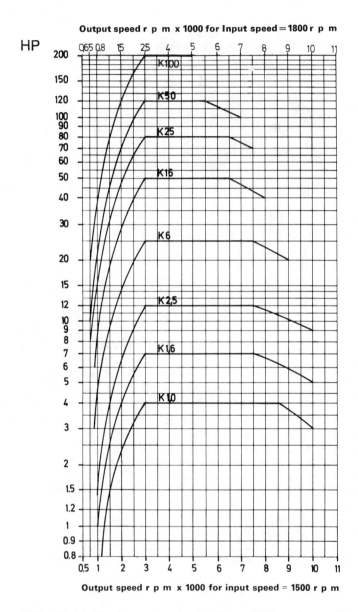

FIG. 12.12 Performance chart, Kopp KHS series. (Courtesy of Kopp-Variatoren AG, Mürten, Switzerland.)

12.7 ALTERNATOR DRIVE FOR SOLAR ENERGY PLANT [6]

The Ottobrunn Division of Messerschmitt-Bolkom-Blohm (MBB), with the support of the Ministry for Research and Development of the Federal Republic of Germany, has built a solar energy system for the generation of electricity. The 10-kW power plant is designed for use by a small settlement for power to be used to supply drinking water and irrigation water, to power a communications system and a small workshop, and for lighting and air conditioning.

One of the first units delivered was set up in India. The alternator is driven at constant speed (1500 rpm) by a Kopp type HS variator. This drive was chosen because it was capable of operating at high input speed, up to 7500 rpm, and exhibits high efficiency despite a high-temperature environment.

Figure 12.13 is the schematic diagram of the power loop in the plant. Solar energy, trapped by the collectors (1), heats a transfer fluid which is pumped through a thermal energy storage unit (2). Hot water is pumped (3) in a closed loop, picking up heat from the storage unit and delivering it to an evaporator (4), where it gives up heat to a working fluid and then is returned to the thermal energy storage unit to pick up more heat.

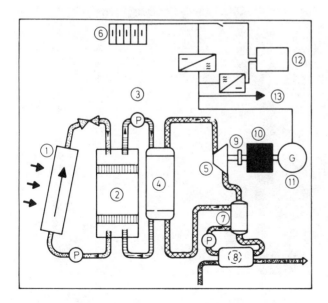

FIG. 12.13 Schematic of the MBB solar energy plant. (Courtesy of Kopp-Variatoren AG, Mürten, Switzerland.)

In the evaporator, a Rankine cycle working fluid is flashed into vapor at high pressure, driving a screw expander (5). The speed of the expander can be 1500 to 7500 rpm, depending on the amount of energy available. Leaving the expander at low pressure, the vapor passes through a recuperator (7), where some of its energy is used to heat to liquid passing from the condenser (8) to the evaporator. The vapor then goes to the condenser, where it is liquefied at low pressure by cooling water, raised to high pressure by a pump, and returned to the evaporator via the recuperator, completing the cycle.

The variable speed of the expander is converted to constant 1500-rpm alternator speed by the traction drive, a Kopp K6HS, which is servo controlled to maintain constant output speed. The output shaft drives an alternator which furnishes 2.5 to 10 kW, depending on the amount of solar energy available.

Some of the alternator output is used by the loads (13) described earlier in this section. Some alternator output is rectified and stored in batteries (6) for use by the pump motors and controls (2).

Figure 12.14 is a photograph of the machinery used in this installation. The variator is in the center of the photograph. The screw expander is on the left, within the frame. The alternator is on the

FIG. 12.14 Traction drive installation for solar energy plant. (Courtesy of Kopp-Variatoren AG, Mürten, Switzerland.)

FIG. 12.15 Solar energy plant. (Courtesy of Kopp-Variatoren AG, Mürten, Switzerland.)

right. Two of the pumps can be seen in the bottom part of the figure. Figure 12.15 is a general view of the installation site. Note the prominence of the thermal storage unit and the solar energy collectors compared to the size of the machinery building.

12.8 MACHINE TOOLS

Material removal with a machine tool requires two essential motions, feed and speed. The machine on which the material removal takes place can be characterized by which element has the feed and which has the speed. In a drill press the tool has both the feed and the speed. In a lathe the workpiece has the speed and the tool has the feed. In a milling machine the workpiece has the feed and the tool has the speed.

The feed, speed, and tool shape depend on the material being cut. Wherever feed or speed have to be varied to suit operating conditions, applications for traction drives will be found. Traction drives are not the only way in which required feeds and speeds may be set up. Change gearing, stepped pulleys, and electronic and hydraulic drives have satisfied a majority of the applications.

Following are several examples of successful traction drive applications in the machine tool industry.

1. Copy lathe [1]. The Dean Smith & Grace [7] type 13/1 lathe has a 7.5-hp motor driving the headstock through an F.U. Type V9 traction drive, which has a 6:1 speed range. The headstock nominal speed is established by the change gears. The lathe has a copy attachment which enables it to trace various internal and external contours.

The part diagrammed in Fig. 12.16 requires a cutting speed no greater than 320 ft min^{-1} to achieve the required finish. This could

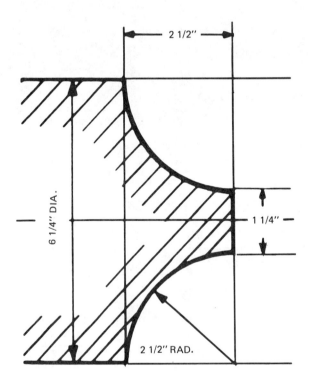

FIG. 12.16 Typical detail part made on a copy-turning lathe. (Courtesy of F.U. (London) Ltd., Leicester, England.)

Machine Tools 311

FIG. 12.17 Traction drive to vary headstock speed of copy lathe. (Courtesy of F.U. (London) Ltd., Leicester, England.)

be done at a constant headstock speed of 194 rpm, established by the outside diameter of the part. The time for the necessary finish cut would be 77 sec.

If the part were finished at constant surface speed, the headstock would have to speed up as the tool traced the smaller-diameter areas. Final speed would be 840 rpm as the tool traced out the smallest diameter. Time for the finish cut would be 40 sec, a saving of 48%.

Figure 12.17 shows the traction drive that accomplishes the headstock speed change as a function of the tool position.

2. Boring machine [6]. Figure 12.18 shows a two-head boring machine (Zehnder) with a Kopp K2.5 variator and 3-kW (4-hp) motor on each Chiron boring head. The variator input and output shafts are on the same side of the drive. The speed range of each head is 330 to 2300 rpm. This machine drills 40 mm (1.575 in.) diameter in cast iron and 30 mm (1.181) diameter in steel, and bores up to 250 mm (9.843 in.) diameter. The feed is infinitely variable under hydraulic control.

3. Milling machine [6]. Figure 12.19 shows a milling machine (Schaffner) with a 3-kW (4-hp) motor driving the main spindle through a Kopp K2.5 variator. Note the speed-adjustment knob on a flexible cable extension on the right side of the spindle. This replaces the usual array of change levers and pulleys. The spindle speed is adjustable from 400 to 1800 rpm.

4. Stamping press [6]. This application is similar to that described in some detail in Section 12.1. This Feintool stamping press (Fig. 12.20) is driven by a 15-kW (20-hp) motor through a Kopp K16

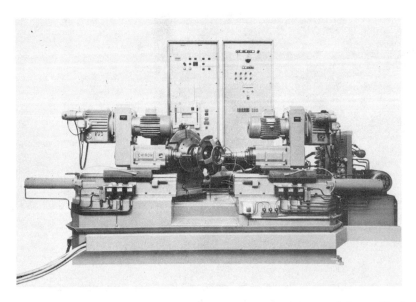

FIG. 12.18 Boring machine (Zehnder). (Courtesy of Kopp-Variatoren AG, Mürten, Switzerland.)

variator which allows the punch to operate over a range of 18 to 72 strokes per minute. The original design was rated at 100 tons (224,000 lb force), but after service experience with the drive the rating was increased to 160 tons (358,400 lb).

    5. Grinding machine [8]. See Fig. 12.21. The traction drive is operating the work spindle.
    6. Thread-cutting machines [8]. See Fig. 12.22.
    7. Small lathe headstock drive [9]. See Fig. 12.23.

12.9 OTHER INDUSTRIAL APPLICATIONS

    1. Synthetic-fiber-spinning equipment [8]. See Fig. 12.24.
    2. Circular annealing conveyors [10]. See Fig. 12.25.
    3. Compactor feed [10]. See Fig. 12.26.
    4. Chipboard-making machine [8]. See Fig. 12.27. The traction drive is operating the chip mechanism at the lower left.
    5. Input nip roll drive for a granulator [8]. See Fig. 12.28.

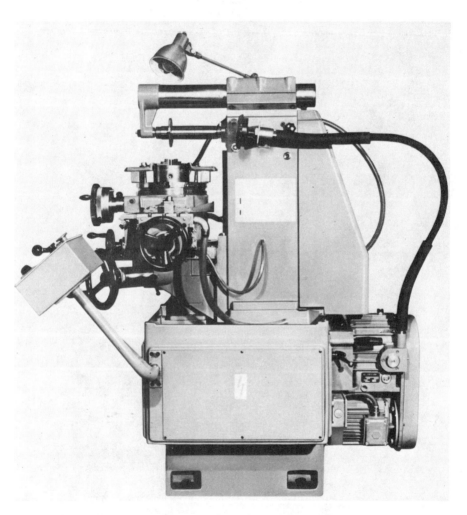

FIG. 12.19 Milling machine (Schaffner). (Courtesy of Kopp-Variatoren AG, Mürten, Switzerland.)

FIG. 12.20 Stamping press (Feintool). (Courtesy of Kopp-Variatoren AG, Mürten, Switzerland.)

FIG. 12.21 Grinding machine with traction drive headstock. (Courtesy of P.I.V. Antrieb Werner Reimers, Bad Homburg, Germany.)

FIG. 12.22 Thread-cutting machines. (Courtesy of P.I.V. Antrieb Werner Reimers, Bad Homburg, Germany.)

FIG. 12.23 Variable headstock drive for small lathe. (Courtesy of Walter V. Chery, Meadville, Pa.)

12.10 ADDITIONAL APPLICATIONS

A request to traction drive suppliers for applications to use in this chapter brought many more than can be described in detail here. Other uses served successfully by traction drives include:

    Band saws
    Blowers
    Bottle conveyers and filling machinery
    Can seam welding machine
    Coating and embossing rollers
    Conveyer pay-off reels
    Drawing machines
    Driers
    Drill presses
    Extruders
    Fans
    Feeders
    Film processors

FIG. 12.24 Synthetic-fiber-spinning equipment. (Courtesy of P.I.V. Antrieb Werner Reimers, Bad Homburg, Germany.)

FIG. 12.25 Circular annealing conveyors. (Courtesy of Simplana Corporation, Ann Arbor, Mich.)

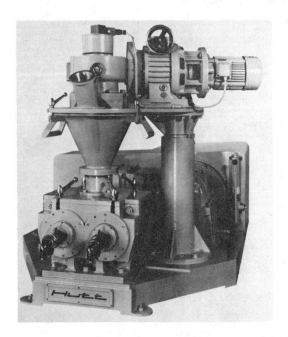

FIG. 12.26 Compactor feed. (Courtesy of Simplana Corporation, Ann Arbor, Mich.)

    Labeling equipment
    Looms
    Mixers
    Pumps; metering pumps
    Sheeters
    Stirrers
    Winders
    Wire-wrap machines

In general, traction drives have a special place when ratio needs to be changed during operation in a stepless, infinitely variable manner, and when the resulting new ratio has to be held reasonably constant until next changed.

This is not to imply that variable-ratio drives are the only traction applications. Wedgtrac Corporation [11] manufactures constant-ratio drives used competitively with geared drives. The Wedgtrac drives are said to be quiet, efficient, and cost competitive with geared speed reducers. In one application in which the driven machine would

FIG. 12.27 Chipboard-making machine. (Courtesy of P.I.V. Antrieb Werner Reimers, Bad Homburg, Germany.)

FIG. 12.28 Drive for the input nip rolls of a granulator. (Courtesy of P.I.V. Antrieb Werner Reimers, Bad Homburg, Germany.)

be damaged if accidentally reversed due to miswiring, the Wedgtrac drive was selected because the wedge roller disengages and the drive uncouples if reversed.

The late A. L. Nasvytis of Nastec Inc. [12] developed some multiroller fixed-ratio planetary transmissions suitable for 3000-hp helicopter applications.

Paxton Products [13] developed a fixed-ratio planetary drive that is incorporated into a supercharger/blower to enable the impeller to operate at high speed with moderate shaft speed input.

Considerable development is now taking place at many development laboratories throughout the world on the continuously variable transmission (CVT) for automotive use. This takes us back to Chapter 2, where the history of traction was reported. Traction began with automotive applications, but was frustrated by the lack of appropriate materials and lubricants and by relatively high manufacturing costs. Traction has now emerged as a viable competitor for many industrial applications, especially where high speed, high power, and high efficiency are required at once. Since these are the characteristics required for the CVTs of today, we may yet share in the vindication of the early pioneers.

REFERENCES

1. Personal communication, July 3, 1982, Apr. 7, 1982, and Mar. 17, 1982, from Geoffry P. McLeod, Director, F.U. (London) Ltd. (See Appendix A under "Ball Disc" for the addresses of F.U. (London) Ltd. and Parker Industries, Inc.)
2. Personal communication, Feb. 11, 1982, from John Fischer, P.E., Senior Design Engineer, Koppers Company, Inc., Sprout Waldron Division, Muncy, Pa. (See Appendix A under "Kopp Variator" for the home office address of the Koppers Company, Inc., and for more catalog information about the variator.)
3. Specifications for this project submitted by John Fischer, P.E., Senior Design Engineer, Koppers Company, Inc., Sprout Waldron Division, Muncy, Pa.
4. Personal communication, Sept. 23, 1981, from E. C. Tibbals, President, TEK Products Corp. (See Appendix A under "Free Ball" for the company's address.)
5. "Double Production Without Increasing Energy Consumption by Using a Mechanical Speed Variator," Technical Note, submitted by Gene Danowski, Manager of Engineering, Sumitomo Machinery Corporation of America. (See Appendix A under "Disc" for the company's address.)
6. Personal communication, Jan. 22, 1982 and Nov. 10, 1981, from Jean E. Kopp, Kopp-Variatoren AG. (See Appendix A under "Kopp Variators" for the company's address.)
7. "Machine Ship and Engineering Manufacture. Constant Surface Speeds and Horsepower Can Reduce Cutting Times by 50%," advertising fly sheet F.U. (London) Ltd. [1]. (Further information can be obtained from Dean Smith Grace Ltd., P. O. Box 15, Keighley, Yorkshire, U.K.)
8. Personal communication, Feb. 24, 1981, from G. Versock, P.I.V. Antreib Werner Reimers. (See Appendix A under "Ball Disc" for the company's address.)
9. Personal communication, Jan. 26, 1982, from Walter Chery. Reference is also made to "infinitely variable speed drive" by W. Chery, personal communication, Mar. 25, 1977. (See Appendix A under "Planetary, Adjustable Ratio," for Mr. Chery's address.
10. Personal communication, June 9, 1981, from Jeffry C. Comstock, Sales Engineer, Simplana Corporation. (See Appendix A under "Disc" for the company's address.

11. Personal communication, Bertel S. Nelson, President, Wedgtrac Corporation. (See Appendix A under "Planetary, Fixed Ratio" for the company's address.)
12. See Appendix A under "Planetary, Fixed Ratio," for address. (For a technical bibliography see the articles and reports by Loewenthal in Appendix B.)
13. Personal communication, Apr. 6, 1982, from Daniel J. Vander Pyl, Paxton Products. (See Appendix A under "Planetary, Fixed Ratio" for the company's address.)

appendix A
TRACTION DRIVE SOURCES

As with any book, it has been the intention to whet the appetite of readers enough so that they will want to pursue the subject further on their own.

The following is a compilation of the names, addresses, and catalogs of the manufacturers, licensees, and agents of the various traction drives according to their generic type, in the order described in Chapter 3. Although some of the sources listed may not have catalogs available, many have written technical papers on their drives. Inquiries can be made of these sources for copies or where to obtain them. Known sources with papers have been indicated, but no attempt has been made to list the papers.

KOPP VARIATOR

Ball

Allspeeds Limited; P. O. Box 43 Royal Works; Accrington, Lancashire, England
    Technical paper
    Catalog

Cleveland Gear; 3249 East 80th St.; Cleveland, Ohio 44104
    Cleveland Speed Variator Catalog K-206A and Bulletin K-276

Kopp-Variatoren AG or Jean E. Kopp Variators; 3280 Meyriez/Mürten, Switzerland
    Kopp Variators Type M Bulletin M 77

Koppers Company, Inc.; Box 1696; Baltimore, Maryland 21203
    Kopp Variators Type M Bulletin M 77

Winsmith; Division of UMC Industries, Inc.; Springville, New York 14141
Winkopp Adjustable Speed Drives Catalog 323A

Roller

Kopp-Variatoren AG or Jean E. Kopp Variators; 3280 Meyriez/Mürten, Switzerland
K Type Variator Serie 4, Bulletin K 76

Koppers Company, Inc.; Box 1696; Baltimore, Maryland 21203
Technical paper
Koppers Series "R" Adjustable Speed Drive Form No. 7583-20 M Wolk 7/79

BALL DISC

A. C. Compacting Presses, Inc.; P. O. Box 1766; 1577 Livingston Ave.; New Brunswick, New Jersey 08902 (U.S. Agent for P.I.V. Antrieb Werner Reimers)
Variable Speed Drives System KS 165/7

F.U. (London) Ltd.; Martin House; Gloucester Cresent; Wigston, Leicester, England LE8 2YL
FU Variable Speed Units, 4th Edition

Graham Transmissions, Inc.; Division of Stowell Industries, Inc.; Box 160; Menomonee Falls, Wisconsin 53051
Variable Speed Ball/Disc Drive BD-300
Metallic Traction Drive Ball/Disc Series Catalog 4003

Parker Industries, Inc.; 1615 Ninth Ave.; Bohemia, New York 11716 (U.S. agent for the F.U. Unicum variable-speed drives)
Variable Speed Traction Drives Bulletin FU-1

P.I.V. Antrieb Werner Reimers KG; Postfach 1960; D-6380 Bad Homburg, West Germany
Technical paper
Variable Speed Drives System KS 165/7

P. T. Components, Inc.; 2045 W. Hunting Park Ave.; Philadelphia, Pennsylvania 19140 (U.S. agent for P.I.V. Antrieb Werner Reimers)

Rolling Contact Gear Company; Attn: Mr. William Rouverol, President;
1521-A Shattuck Ave.; P. O. Box 9122; Berkeley, California
94709 (inventor of ball disc but does not manufacture the drive)
Technical papers

Unicum S.A.; 35, rue de Bienfaisance; 75008 Paris, France
Catalog

RING CONE

Andantex U.S.A. Inc.; 1800 Brielle Ave.; Ocean, New Jersey 07712
(U.S. agent for Hans Heynau)

Graham Transmissions, Inc.; Division of Stowell Industries, Inc.;
Box 160; Menomonee Falls, Wisconsin 53051
R/C Series Metallic Traction Drives Catalog 9201
R/R Series Metallic Traction Drives Catalog 7004
Ring Cone RX Metallic Traction Drive Catalog 9300

Hans Heynau GmbH; Postfach 40 08 48; D-8000 München 40, West
Germany
H Drive Infinitely Variable 5.0/10.75e

Shimpo American Corp.; 3510 Devon Ave.; Lincolnwood, Illinois 60659
(Graham sells the Shimpo drives in the United States)

Shimpo Industrial Co., Ltd.; 338 Tonoshiro-cho, Kuze, Minami-ku;
Kyoto, Japan 601
Ringcone Variable Speed Drives Catalog 710-A, Dec. 1971

Vadetec Corporation; 2681 Industrial Row; Troy, Michigan 48048
Technical papers

FREE BALL

Andantex U.S.A. Inc.; 1800 Brielle Ave.; Ocean, New Jersey 07712
(U.S. agent for Hans Heynau)

Contraves AG; Postfach Schaffhauserstrasse 580; CH-8052 Zurich,
Switzerland
Technical paper
Continuous Speed Variators Bulletin A 102e-7603
Speed Variators Type 0-1000 Bulletin A 103e-7710

Floyd Drives Company; 3080 Valmont Ave., No. 7; Boulder, Colorado 80301

Hans Heynau GmbH; Postfach 40 08 48; D-8000 München 40, West Germany
Minidrive Catalog 6652/3569D 3.-4.81-SD
Maxi-Minidrive Catalog 2.0/5.77e (discontinued)

TEK Products Corp.; P.O. Box 1652; High Point, N.C. 27261 (U.S. agent for Contraves until late 1970s; now selling just units in existing inventory)

DISC

Maschinenfabrik Hans Lenze KG; Postfach 100; D-4923 Extertall, West Germany
Lenze Power Transmission 10.78.1

Shimpo American Corp.; 3510 Devon Ave.; Lincolnwood, Illinois 60659

Shimpo Industrial Co., Ltd.; 338 Tonoshiro-cho, Kuze, Minami-ku; Kyoto, Japan 601
Ringcone Variable Speed Drives Catalog 710-A, Dec. 1971

Simplana Corporation; P. O. Box 2446; Ann Arbor, Michigan 48106 (U.S. agent for Hans Lenze KG)
Disco Planetary Variable Speed Drives 5/1/80-4987

Sumitomo Machinery Corporation of America; 7 Malcolm Ave.; Teterboro, New Jersey 07608
Technical Paper
SM-Cyclo Beier Variators Bulletin B500A 3/80 2.5M

TOROIDAL

Arter Regelgetriebe Ltd.; Bahngasse 5; CH-8708 Männedorf ZH, Switzerland
Arter Variatoren, Aug. 1976
Infinitely Variable Speed Control with Arter
Traction drives pamphlet
Technical paper

Planetary

David Brown Gear Industries, Inc.; 60 Emblem Court; Agincourt, Ontario, Canada M1S 1B1
   Series 7 Sadivar Mechanical Speed Variators and Variable Speed Drives Publication E.492.12

David Brown Sadi SA; 4 rue des Carburants; B-1190 Bruxelles, Belgium
   Series 7 Sadivar Mechanical Speed Variators and Variable Speed Drives Publication E.492.12

Excelermatic, Inc.; 913 Sagebrush Drive; Austin, Texas 78758
   Technical papers

PLANETARY

Fixed Ratio

NASA Lewis Research Center; Cleveland, Ohio 44135 (not a manufacturer, but conducted extensive testing on the Nasvytis drive)
   Technical papers

Nastec Inc.; 1700 Ohio Savings Plaza; 1801 East 9th St.; Cleveland, Ohio 44114
   Nasvytrac Power Transfer Through Traction pamphlet

Paxton Products; 929 Olympic Blvd.; Santa Monica, California 90404
   Paxton Centrifugal Blowers pamphlet

Traction Propulsion, Inc.; Subsidiary of Excelermatic, Inc.; 913 Sagebrush Drive; Austin, Texas 78758
   Technical paper

Wedgtrac Corporation; Fox Industrial Park; Beaver and Deer Streets; Yorkville, Illinois 60560
   Wedgtrac Drives Form 773-5000

Adjustable Ratio

Mr. Walter Chery; 744 Alden St.; Meadville, Pennsylvania 16335 (inventor of the Chery drive; not actively manufacturing)
   Technical paper

Fafnir Bearing; Division of Textron; 37 Booth Street; New Britain Connecticut 06050 (manufactures a variation of the Chery drive)
   Technical paper

SPOOL

Traction Propulsion, Inc.; Subsidiary of Excelermatic, Inc.; 913 Sagebrush Drive; Austin, Texas 78758 (limited number built for special applications)
Technical paper

OFFSET SPHERE

J. R. Young Corp.; 4956 Douglas Ave.; Racine, Wisconsin 53402 (last known address, but mail not deliverable and presumed no longer in business)

FRICTION DRIVE

Eurodrive Inc.; 2001 West Main Street; Troy, Ohio 45373
    Varimot Catalog E-1235-58010M and Flysheet E4-5802-5M11
    Technical paper
McDonough Power Equipment; McDonough, Georgia 30253

appendix B

BIBLIOGRAPHY

Algayer, B., "Future Trends in Mechanical Variable Speed Drives," Philadelphia Gear, King of Prussia, Pa., Nov. 1974.

Appleton's Cyclopedia of Applied Mechanics, Vol. II, Part D, Appleton, New York, 1880, pp. 36-37.

Bamberger, E. N., "Materials for Rolling Element Bearings. Bearing Design—Historical Aspects, Present Technology, and Future Problems," American Society of Mechanical Engineers, New York, 1980, pp. 1-46.

Bamberger, E. N., et al., "Life Adjustment Factors for Ball and Roller Bearings—An Engineering Guide," American Society of Mechanical Engineers, New York, 1971.

"Basic Types of Traction Drives," Mechanical Drives Reference Issue, *Mach. Des.*, June 19, 1980.

Beachley, N. H., and Frank, A. A., "Increased Fuel Economy in Transportation Systems by the Use of Energy Management," University of Wisconsin Report to DOT, DOT-TST-75-2, 1975.

Beachley, N. H., and Frank, A. A., "Principles and Definitions for Continuously Variable Transmissions, with Emphasis on Automotive Applications," ASME Technical Paper 80-C2/DET-95, 1980.

Bell, J. C., Kannel, J. W., and Allen, C. M., "The Rheological Behavior of the Lubricant in the Contact Zone of a Rolling Contact System," *J. Basic Eng.*, vol. 83, no. 3, Sept. 1964, pp. 423-425.

Bhise, A. S., "The Winkopp, Infinitely Variable Speed Traction Drive," National Conference on Power Transmission Transactions, 1979, pp. 255-258.

Blair, S., and Winer, W. O., "Shear Strength Measurements of Lubricants at High Pressures," *J. Lubr. Technol.*, vol. 101, no. 3, July 1979, pp. 251-257.

Cahn-Speyer, P., "Mechanical Infinitely Variable Speed Drives," *Eng. Dig. (Lond.)*, vol. 18, no. 2, Feb. 1957, pp. 41-43.

Caris, D. F., and Richardson, R. A., "Engine-Transmission Relationships for High Efficiency," *SAE Trans.*, vol. 61, 1953, pp. 81-96.

Carson, R. W., "Focus on Traction Drives," *Power Transm. Des.*, vol. 17, no. 3, Mar. 1975, pp. 48-49.

Carson, R. W., "Focus on Traction Drives," *Power Transm. Des.*, vol. 17, no. 8, Aug. 1975.

Carson, R. W., "Focus on Traction Drives: 100 Years of Traction Drives," *Power Transm. Des.*, vol. 17, no. 5, May 1975, pp. 84, 88.

Carson, R. W., "New and Better Traction Drives Are Here," *Mach. Des.*, vol. 46, no. 10, Apr. 18, 1974, pp. 148-155.

Carson, R. W., "New Lubricant Doubles Life of Rolling Element Bearings," *Des. News*, Oct. 6, 1975.

Carson, R. W., "100 Years in Review: Industrial Traction Drives," *Power Transm. Des.*, vol. 19, no. 10, Oct. 1977, pp. 99-100.

Carson, R. W., "Today's Traction Drives," *Power Transm. Des.*, vol. 17, no. 11, Nov. 1975, pp. 41-49.

Carson, R. W., "Traction Drives Update," *Power Transm. Des.*, vol. 19, no. 11, Nov. 1977, pp. 37-42.

Carter, F. W., "On the Action of a Locomotive Driving Wheel," *Proc. R. Soc. (Lond.) A*, vol. 112, 1926, pp. 151-157.

Cheng, H. S., and Sternlicht, B., "A Numerical Solution for the Pressure, Temperature, and Film Thickness Between Two Infinitely Long Lubricated Rolling and Sliding Cylinders Under Heavy Loads," *J. Basic Eng.*, vol. 87, no. 3, Sept. 1965, pp. 695-707.

Chow, T. S., and Saibel, E., "The Elastohydrodynamic Problem with a Viscoelastic Fluid," *J. Lubr. Technol.*, Jan. 1971, pp. 25-31.

Clark, O. H., Woods, W. W., and White, J. R., "Lubrication at Extreme Pressure with Mineral Oil Films," *J. Appl. Phys.*, vol. 22, no. 4, Apr. 1951, pp. 474-483.

Clymer, F., *Historical Motor Scrapbook*, Vols. 1 and 2, Clymer Motor Publications, Los Angeles, 1944.

Bibliography

"Continuously Variable Transmission," *Automot. Eng. Mag.*, vol. 83, no. 12, Dec. 1975.

Coy, J. J., Rohn, D. A., and Loewenthal, S. H., "Constrained Fatigue Life Optimization of a Nasvytis Multiroller Traction Drive," NASA Technical Memorandum 81447, AVRADCOM Technical Report 80-C-6, Aug. 1980.

Coy, J. J., Rohn, D. A., and Loewenthal, S. H., "Life Analysis of Multiroller Planetary Traction Drive," NASA Technical Paper 1710, AVRADCOM Technical Report 80-C-16, Apr. 1981.

Coy, J. J., Loewenthal, S. H., and Zaretsky, E. V., "Fatigue Life Analysis for Traction Drives with Application to a Toroidal Type Geometry," NASA TN D-8362, 1976.

Crook, A. W., "Lubrication of Rollers. Part IV. Measurements of Friction and Effective Viscosity," *Philos. Trans. R. Soc. (Lond.) A*, vol. 255, no. 1056, Jan. 1963, pp. 281-312.

Culp, D. V., Stover, A., and Stover, J. D., "Bearing Fatigue Life Tests in a Synthetic Traction Lubricant," ASLE Preprint 75AM-1B-1, May 1975.

Daniels, B. K., "Traction Contact Optimization," ASLE Preprint 79-LC-1A-1, 1979.

Dawson, D., and Higgmoon, G. R., *Elastohydrodynamic Lubrication*, Pergamon Press, Elmsford, N.Y., 1966.

DeBono, E., *Eureka: An Illustrated History of Inventions from the Wheel to the Computer*, Holt, Rinehart and Winston, New York, 1974, p. 33.

Dickinson, T. W., "The Development of a Continuously Variable Ratio Transmission for Mobile Equipment," Fafnir Bearing, Division of Textron, New Britain, Conn.

Dickinson, T. W., "Development of a Variable Speed Transmission for Light Tractors," SAE Paper 770749, Sept. 1977.

Dickinson, T. W., "A Variable Speed Traction Transmission for Light Tractors," Fafnir Bearing, Division of Textron, New Britain, Conn.

Dowson, D., and Whitaker, B. A., "A Numerical Procedure for the Solution of the Elastohydrodynamic Problem of Rolling and Sliding Contacts Lubricated by a Newtonian Fluid," *Proc. Inst. Mech. Eng. (Lond.)*, vol. 180, pt. 3B, 1965, pp. 57-71.

Drachmann, A. G., *The Mechanical Technology of Greek and Roman Antiquity*, Lubrecht and Cramer, 1963, pp. 200-203.

Dudley, D. W., "The Evaluation of the Gear Art," American Gear Manufacturers Association, Washington, D.C., 1969.

Dvorak, D. Z., "Your Guide to Variable-Speed Mechanical Drives," *Prod. Eng.*, vol. 34, Dec. 1963, pp. 63-74.

Dyson, A., "Frictional Traction and Lubricant Rheology in Elastohydrodynamic Lubrication," *Phil. Trans. R. Soc. (Lond.) A*, vol. 266, no. 1170, Feb. 1970, pp. 1-33.

Eckert, F. W., "Koppers Series 'R' Adjustable Speed Drive," Technical Paper, National Conference on Power Transmission Transactions, 1975.

Edsall, B., "The Ideal Automatic Transmission," *Automot. Ind.*, May 1953.

Edson, D. V., "Continuously Variable Transmission: a Concept Whose Time Has Come?" *Des. News*, Jan. 5, 1981, pp. 42-55.

Elfes, L. E., "Fixed and Adjustable Speed Drives," National Conference on Power Transmission Transactions, 1981.

Elu, P., and Kemper, Y., "Performance of a Nutating Traction Drive," ASME Paper 80-C2/DET-63, Aug. 1980.

Ertel, A. M., "Hydrodynamic Lubrications Based on New Principles," *Prikl. Mat. Mekh.*, vol. 3, no. 2, 1939 (in Russian).

Fellows, T. G., et al., "Perbury Continuously Variable Ratio Transmission," in *Advances in Automobile Engineering, Part II*, Pergamon Press, Oxford, 1964, pp. 123-142.

"Final Report on the Follow-on Phase of the Evaluation of the Wright Aeronautical Division Toroidal Drive Using a $\frac{1}{4}$ Ton Military Truck as a Test Bed Vehicle," WAD R273-F, Curtis-Wright Corporation, Wood Ridge, N.J., Feb. 1966.

Foord, C. A., Hammann, W. C., and Cameron, A., "Evaluation of Lubricants Using Optical Elastohydrodynamics," *ASLE Trans.*, vol. 11, 1968, pp. 31-43.

Gaggermeier, H., "Investigations of Tractive Force Transmission in Variable Traction Drives in the Area of Elastohydrodynamic Lubrication," Ph.D. dissertation, Technical University of Munich, July 1977.

Green, R. L., and Langenfeld, F. L., "Lubricants for Traction Drives," *Mach. Des.*, May 2, 1974.

Grubin, A. N., "Fundamentals of the Hydrodynamic Theory of Lubrication of Heavily Loaded Cylindrical Surfaces. Investigation of

the Contact of Machine Components," in Kh. F. Ketova, ed.,
Translation of Russian Book No. 30, Central Scientific Institute
for Technology and Mechanical Engineering, Moscow, 1949,
Chap. 2. (Available from Department of Science and Industrial
Research, Great Britain, Transl. CTS-235 and Special Libraries
Association, Transl. R-3554.)

Hamman, W. C., et al., "Synthetic Fluids for High Capacity Traction
Drives," *ASLE Trans.*, vol. 13, 1970, pp. 105-116.

Hamrock, B., and Dowson, D. *Ball Bearing Lubrication*, Wiley,
New York, 1981.

Hazeltine, M. W., "Rolling Contact Slip or Wear? Traction Fluid Takes
the Heat Off," *Power Transm. Des.*, vol. 15, no. 11, Nov. 1973,
pp. 64-68.

Hazeltine, M. W., et al., "Design and Development of Fluids for Traction
and Friction Type Transmissions," SAE Paper 710837, Oct. 1971.

Heldt, P. M., "Automatic Transmissions," *SAE J.*, vol. 40, no. 5,
May 1937, pp. 206-220.

Hewko, L. O., "Contact Traction and Creep of Lubricated Cylindrical
Rolling Elements at Very High Surface Speeds," *ASLE Trans.*,
vol. 12, no. 2, Apr. 1969, pp. 151-161.

Hewko, L. O., "Roller Traction Drive Unit for Extremely Quiet Power
Transmission," *J. Hydronaut.*, vol. 2, no. 3, July 1968, pp.
160-167.

Hewko, L. O., "Traction Drives and Their Potential Role in Energy
Conservation," Joint ASLE Energy-Sources Technology Conference, New Orleans, La., Feb. 1980.

Hewko, L. O., Rounds, F. G., Jr., and Scott, R. L., "Tractive
Capacity and Efficiency of Rolling Contacts," in *Rolling Contact
Phenomena*, J. B. Bidwell, ed., Elsevier, Amsterdam, 1962, pp.
157-185.

Hitchcox, A., "Sorting Out Traction Drives," *Power Transm. Des.*,
vol. 23, no. 8, Aug. 1981, pp. 51-53.

Hodges, D., and Wise, D. B., *The Story of the Car*, Hamlyn, London,
1974, p. 18.

Homans, J. E., *Self Propelled Vehicles, a Practical Illustrated Treatise
on Automobiles*, Theodore Audel, New York, 1910 rev. ed., pp.
39, 361-362, 364-365.

"Improved Technology Is Giving an Old Principle a New Drive,"
Monsanto Mag., Summer 1974, pp. 14-16.

Johnson, K. L., "Tangential Tractions and Micro-slip in Rolling Contact," in *Rolling Contact Phenomena*, J. B. Bidwell, ed., Elsevier Publishing Co., Amsterdam, 1962, pp. 6-28.

Johnson, K. L., and Cameron, R., "Shear Behavior of Elastohydrodynamic Oil Films at High Rolling Contact Pressures," *Proc. Inst. Mech. Eng. (Lond.)*, vol. 182, no. 1, 1967, pp. 307-318.

Johnson, K. L., and Greenwood, J. A., "Thermal Analysis of an Eyring Fluid in Elastohydrodynamic Traction," *Wear*, vol. 61, 1980, pp. 353-374.

Johnson, K. L., Nayak, L., and Moore, A. J., "Determination of Elastic Shear Modulus of Lubricants from Disc Machine Tests. Elastohydrodynamics and Related Topics," *Proc. 5th Leeds-Lyon Symposium on Tribology*, Mechanical Engineering Publications, Suffolk, England, 1979, pp. 204-213.

Johnson, K. L., and Roberts, A. D., "Observation of Visco Elastic Behavior of an Elastohydrodynamic Lubricant Film," *Proc. R. Soc. (Lond.) A*, vol. 337, 1974, pp. 217-242.

Johnson, K. L., and Tevaarwerk, J. L., "Shear Behavior of Elastohydrodynamic Oil Films," *Proc. R. Soc. (Lond.) A*, vol. 356, no. 1685, Aug. 1977, pp. 215-236.

Kacmarsky, W. M., and Hewko, L. O., "Effect of a High Traction Fluid on Skidding in a High Speed Roller Bearing," *J. Lubr. Technol.*, Jan. 1971, pp. 11-16.

Kannel, J. W., and Walowit, J. A., "Simplified Analysis for Tractions Between Rolling-Sliding Elastohydrodynamic Contacts," *J. Lubr. Technol.*, vol. 93, Jan. 1971, pp. 39-46.

Kemper, Y., "A High Power Density Traction Drive," SAE Paper 790849, 1979.

Kemper, Y., and Elfes, L., "A Continuously Variable Traction Drive for Heavy Duty Agricultural and Industrial Applications," SAE Paper, Jan. 1981.

Kirkwood, T. F., and Lee, A. D., "A Generalized Model for Comparing Automobile Design Approaches to Improve Fuel Economy," Report to National Science Foundation R-1562-NSF, Jan. 1975.

Knights, E. H., *Knights' American Mechanical Dictionary*, Vol. I, Hurd and Houghton, 1876, p. 680.

Kraus, C. E., "New Approaches to Variable-Speed Drives," *Mach. Des.*, Dec. 1953, pp. 232-246.

Kraus, C. E., "Rolling Traction Analysis and Design," Excelermatic, Inc., Austin, Tex., 1972.

Kraus, C. E., "Traction Drives for Vehicles": "Part I: The Ideal Drive," "Part II: Matching the Ideal," "Part III: Recent Advances," *Power Transm. Des.*, vol. 10, no. 3, Mar. 1968, pp. 58-61; vol. 10, no. 5, May 1968, pp. 63-66; vol. 10, no. 7, July 1968, pp. 50-53.

Kraus, C. E., "A Transaxle Design for a Traction Continuously Variable Transmission for Automobiles," SAE/SP-80/465, 800102, 1980.

Kraus, C. E., "Understanding Traction," *Power Transm. Des.*, vol. 22, no. 5, May 1980, pp. 36-39.

Kraus, C. E., "An Up-to-Date Guide for Designing Traction Drives," I, II, *Mach. Des.*, July 2, 1964, pp. 106-112; July 16, 1964, pp. 147-152.

Kraus, C. E., "What Does the Future Hold for Industrial Traction Drives?" National Conference on Power Transmission, Oct. 1980.

Kraus, C. E., and Gres, M. E., "Continuously Variable Transmission for Single-Shaft, Turbine Powered Vehicles," ASME Paper 74-GT-106, Mar. 30, 1974.

Kraus, C. E., and Gres, M. E., "New Hope for Single-Shaft Turbine Car," *Gas Turbine Int. Mag.*, May-June 1973.

Kraus, C. E., and Gres, M. E., "A Transmission System for Single-Shaft Gas Turbine Powered Trucks," SAE Paper 730644, June 20, 1973.

Kraus, J. H., "An Automotive CVT," *Mech. Eng.*, Oct. 1976, pp. 38-43.

Kraus, J. H., "High Powered Continuously Variable Traction Drives for Industrial Applications," Excelermatic, Inc., Austin, Tex., Aug. 22, 1977.

Kraus, J. H., "The Selection and Optimization of a Continuously Variable Transmission for Automotive Use," ASME Paper 75-WA/Aut-16, Dec. 2, 1975.

Kraus, J. H., "Traction Drive Shows Automotive Promise," *Mach. Des.*, Oct. 18, 1973.

Kraus, J. H., Kraus, C. E., and Gres, M. E., "A Continuously Variable Transmission for Automotive Fuel Economy," SAE Paper 751180, Jan. 28, 1975.

Kutter, F., "Theoretische Untersuchung eines Reibgetriebes," Thesis, Technical University of Dresden, 1944.

Lane, T. B., "The Lubrication of Friction Drives," *Lubr. Eng.*, vol. 13, Feb. 1957, pp. 85-88.

Lavoie, F. J., "Mechanical Adjustable-Speed Drives," *Mach. Des.*, Sept. 12, 1968.

Lindsley, E. F., "Tracor—Tomorrow's Stepless Transmission for Better Mileage, Faster Starts," *Popular Sci.*, Mar. 1975, pp. 91-93.

Lindsley, E. F., "Traction-Drive Transmission: It's Infinitely Variable, and Helps Recover Braking Energy, Too," *Popular Sci.*, Mar. 1980, pp. 83-86.

Lingard, S., "Tractions at the Spinning Point Contacts of a Variable Ratio Friction Drive," *Tribol. Int.*, vol. 7, Oct. 1974, pp. 228-234.

Lohmann, G., and Schreiber, H. H., "Determination of the Rating Life Exponent for Ball and Roller Bearings," *Werkstatt Betr.*, Zeitschrift für Maschinenban und Festigung, vol. 92, 1959, pp. 188-192.

Loewenthal, S. H., "Advanced Continuously Variable Transmissions for Electric and Hybrid Vehicles," DOE/NASA/51044-17, NASA TM-81718, 1980.

Loewenthal, S. H., "A Historical Perspective of Traction Drives and Related Technology," NASA CP-2210, June 1981.

Loewenthal, S. H., Anderson, N. E., and Nasvytis, A. L., "Performance of a Nasvytis Multiroller Traction Drive," NASA TP-1378, 1978.

Loewenthal, S. H., Anderson, N. E., and Rohn, D. A., "Evaluation of a High Performance Fixed-Ratio Traction Drive," *J. Mech. Des.*, vol. 103, no. 2, Apr. 1981, pp. 410-422.

Loewenthal, S. H., and Parker, R. J., "Rolling-Element Fatigue Life with Two Synthetic Cycloaliphatic Traction Fluids," NASA TN D-8124, 1976.

Lundberg, G., and Palmgren, A., "Dynamic Capacity of Rolling Bearings," Ingenioersvetenskapsakademian Handlinger No. 196, 1947.

Lutz, O., "Grundsatzliches über stufenlos verstellbare Walzgetriebe," *Z. Konstr.*, vol. 7, 1955, p. 330; vol. 9, 1957, p. 169; vol. 10, 1958, p. 425.

Maass, H., "Untersuchung über die in elliptischen Hertzscher Flachen übertragbaren Umfangskrafte," Dissertation, Technical University of Braunschweig, 1959.

MacPherson, P. B., "The Pitting Performance of Hardened Steels," ASME Paper 77-DET-39, Sept. 1977.

Magi, M., "On Efficiencies of Mechanical Coplanar Shaft Power Transmissions," Chalmers University, Gothenburg, Sweden, 1974.

Martin, K. F., "A Review of Frictional Predictions in Gear Teeth," *Wear*, vol. 49, 1978, pp. 201-238.

Martin, M. K., "Advanced Automotive Transmission Development Status and Research Needs," The Aerospace Corporation, for U.S. Department of Energy, Aerospace Report ATR-82(2869)-2ND, 1982.

McCormick, D., "Traction Drives Move to Higher Powers," *Des. Eng.*, Dec. 1980, pp. 35-39.

Meyer, S., and Connelly, R. E., "Traction Drive for Cryogenic Boost Pump," NASA TM-81704, 1981.

Nakamura, L., et al., "A Development of a Traction Roller System for a Gas Turbine Driven APU," SAE Paper 790106, Feb. 1979.

Nasvytis, A. L., "Multiroller Planetary Friction Drives," SAE Paper 660763, Oct. 1966.

"New Departures in Automotive Ideas," New Departure Press, General Motors Corp., 1937, pp. 5-6.

"New Transmission Promises Use of Smaller Engines and Better Mileage," *Motor Trend*, Sept. 1959.

Parker, R. J., Loewenthal, S. H., and Fischer, G. K., "Design Studies of Continuously Variable Transmissions for Electric Vehicles," DOE/NASA/10444-12, NASA TM-81642, 1981.

Perry, F. G., "The Perbury Transmission," ASME Paper 80-GT-22, Mar. 1980.

Plint, M. A., "Traction in Elastohydrodynamic Contacts," *Proc. Inst. Mech. Eng. (Lond.)*, vol. 182, no. 1, 1967, pp. 300-306.

"Potential for Motor Vehicle Fuel Economy Improvements," Report to Congress by DOT and EPA, Oct. 24, 1974.

Poon, S. Y., "Some Calculations to Assess the Effect of Spin on the Tractive Capacity of Rolling Contact Drives," *Proc. Inst. Mech. Eng (Lond.)*, vol. 185, no. 76/71, 1970, pp. 1015-1022.

Poritsky, H., Hewlett, C. W., Jr., and Coleman, R. E., Jr., "Sliding Friction of Ball Bearings of the Pivot Type," *J. Appl. Mech.*, vol. 24, no. 4, Dec. 1947, pp. A-261 to A-268.

Radtke, R. R., Unnewehr, L. E., and Freedman, R. J., "Optimization of a Continuously Variable Transmission with Emission Constraints," SAE Paper 810107, Feb. 1981.

Raynard, A. E., Kraus, J. H., and Bell, D. D., "Design Study of Toroidal Traction CVT for Electric Vehicles," Rep.-80-16762, AiResearch Manufacturing Company; NASA contract.

Redkiewics, C. M., and Srinivasan, V., "Elastohydrodynamic Lubrication in Rolling and Sliding Contacts," *J. Lubr. Technol.*, Oct. 1972, pp. 324-329.

"Ring Metal-to-Metal Drives Are Latest Entries for Adjustable Speed," *Prod. Eng.* (N.Y.), Oct. 1973.

Rohn, D. A., Loewental, S. H., and Coy, J. J., "Simplified Fatigue Life Analysis for Traction Drive Contacts," *J. Mech. Des.*, vol. 103, no. 2, Apr. 1981, pp. 430-439.

Rouverol, W. S., "Inertial Effects in a Multiple-Ball Transmission," *J. Eng. Ind.*, Nov. 1960, pp. 399-406.

Rouverol, W. S., and Tanner, R. I., "A Brief Examination of Factors Affecting Tractive Friction Coefficients of Spheres Rolling on Flat Plates," *ASLE Trans.*, vol. 3, no. 1, pp. 11-17.

Schmidt, P. W., "Graham Ring/Roller Adjustable Speed Drive," National Conference on Power Transmission Transactions, 1979, pp. 247-249.

Schoch, W., "The Speed Variator as a Torque Amplifier," Translated from *Werkstatt Betr.*, vol. 98, Jan. 1965, pp. 14-16.

"Slip-Stick Fluids and Superemulsions Defy Normal Fluid Principles," *Prod. Eng.* (N.Y.), Aug. 1971.

Sloan, A. P., *My Years with General Motors*, Doubleday, New York, 1964.

Smith, F. W., "The Effect of Temperature in Concentrated Contact Lubrication," *ASLE Trans.*, vol. 5, no. 1, Apr. 1962, pp. 142-148.

Soda, N., and Yamamoto, T., "Effect of Tangential Traction and Roughness on Crack Initiation/Propagation During Rolling Contact," NASA TM-81608, 1981.

Soderholm, L., "Traction Lubricant Increases Roller Drive Power Transmission," *Des. News*, Sept. 20, 1971.

"Steel-on-Steel," *Des. News*, Aug. 20, 1973, p. 30.

Stephenson, R. R., "Should We Have a New Engine," Jet Propulsion Laboratory, Sunnyvale, Calif., Aug. 1975.

Strauch, S., "Flywheel Systems for Vehicles," *Proc. Electric and Hybrid Vehicle Advanced Technology Seminar*, California Institute of Technology, Pasadena, Dec. 8-9, 1980, pp. 219-223.

Stubbs, P. W. R., "The Development of a Perbury Traction Transmission for Motor Car Applications," *J. Mech. Des.*, vol. 103, no. 4, Jan. 1981, pp. 29-40.

Stuemky, R. E., "The Gates Roller Planetary Drive," National Conference on Power Transmission Transactions, 1979, pp. 251-253.

Tallian, T. E., Chiu, Y. P., and Van Amerogen, E., "Predictions of Traction and Microgeometry Effects on Rolling Contact Fatigue Life," *J. Lubr. Technol.*, vol. 100, no. 2, Apr. 1978, pp. 156-166.

Tevaarwerk, J. L., "A Simple Thermal Correction for Large Spin Traction Curves," *J. Mech. Des.*, vol. 103, no. 2, Apr. 1981, pp. 440-446.

Tevaarwerk, J. L., "Thermal Influence on the Traction Behavior of an Elastic/Plastic Model," Leeds-Lyon Conference on Tribology, Leeds, England, Aug. 1981.

Tevaarwerk, J. L., "Traction Calculations Using the Shear Plane Hypothesis," in *Thermal Effects in Tribology: Proc. 6th (1979) Leeds-Lyon Conference on Tribology*, Mechanical Engineering Publication, Suffolk, England, 1980, pp. 201-213.

Tevaarwerk, J. L., "Traction Contact Performance Evaluation at High Speeds," NASA CR-165226, 1981.

Tevaarwerk, J. L., "Traction Drive Performance Prediction for the Johnson and Tevaarwerk Traction Model," NASA TP-1530, 1979.

Thomas, W., "Reibscheiben-Regelgetriebe" (Linienberuhrung), Vol. 4 of the *Schriftenreihe Antriebstechnik*, Friedr. Vieweg, Braunschweig, 1954.

Trachman, E. G., and Cheng, H. S., "Thermal and Non-Newtonian Effects on Traction in Elastohydrodynamic Contacts," Symposium on Elastohydrodynamic Lubrication, Institute of Mechanical Engineers, London, 1972, pp. 142-148.

Trobridge, "Census Office Report on Power and Machinery Used by Manufacturer," Washington, D.C., 1888, p. 220.

"Updating an Old Transmission," *Bus. Week Mag.*, Feb. 23, 1976, p. 86M.

Walker, R. D., and McCoin, D. K., "Design Study of a Continuously Variable Cone/Roller Traction Transmission for Electric Vehicles," DOE/NASA/0115-80/1, NASA CR-159841, Sept. 1980.

Walowit, J. A., and Smith, R. O., "Traction Characteristics of a MIL-L-7808 Oil," ASME Paper 76-Lubs-19, 1976.

Warner, S. D., "A Traction Drive Mechanical Variable Speed System," Eurodrive, Inc., Troy, Ohio.

Wernitz, W., "Friction at Hertzian Contact with Combined Roll and Twist," in *Rolling Contact Phenomena*, J. B. Bidwell, ed., Elsevier, Amsterdam, 1962, pp. 132-156.

Wernitz, W., *Friction Drives. Mechanical Design and Systems Handbook*, H. A. Rothbart, ed., McGraw-Hill, New York, 1964, pp. 14-1 to 14-22.

Wernitz, W., "Walz-Bohrreibung-Bestimmung der Bohrmomente und Umfangskrafte bei Hertzscher Pressung mit Punktberuhrung," Vol. 19 of *Schriftenreihe Antriebstechnik*, Friedr Vieweg, Braunschweig, 1958.

"What Is a Cone-Drive Adjustable Speed Drive," *Power Transm. Des.*, vol. 19, no. 12, Dec. 1977, pp. 62-63.

Yeaple, F., "Metal-to-Metal Traction Drives Now Have a New Lease on Life," *Prod. Eng.*, vol. 42, no. 15, Oct. 1971, pp. 33-37.

INDEX

Acceleration, 120, 129-130
  of gravity, 120
Adjustment repeatability, 258
AEG-Regulies Getriebe, 24
Aiding load, 236
Allspeeds Ltd., 26, 28
Alternatives, evaluation of
  economic, 264
Alternator drive, 307
Annuity, 267, 273
  tables, 268-269
Analysis
  contact area, 168-176
  dimensional, 121
Angle
  contact inclination, 163
  traction, 149
Applications
  alternator drive, 307
  cutoff machine, 302-306
  knitting machine, 296-298
  machine tools,
    boring machine, 311
    copy lathe, 310-311
    grinding machine, 312
    lathe headstock drive, 312
    milling machine, 311
    stamping press, 311
    thread-cutting machine, 312

[Applications]
  other industrial applications
    chipboard-making machine, 316
    circular annealing conveyor, 316
    compactor feed, 316
    input nip roll drive, 316
    synthetic-fiber-spinning equipment, 316
    paint mixer, 298-303
    solar energy plant, 307-309
    waste compactor, 290-295
Arter, Jacob, 24
Arter, Martin, 68
Arter Regelgetriebe Ltd., 87
Automobiles
  Ames, 14
  Austin, 15
  Avco, 17
  Buick, 16
  Burns, 11
  Cadillac, 16
  Cannon, 8
  Cartercar, 9
  Duryea, 8
  Earl, 11
  Gearless, 11
  Holmes, 11

[Automobiles]
  International, 11
  Kelsey, 14
  Lambert, 8
  Lewis, 8
  Marble Swift, 9
  Metz, 11
  Pittsburgh, 14
  Pontiac, 11
  Quadricycle, 8
  Rambler, 17
  Sears, 9
  Simplicity, 9
  Stanley, 11
  Stanley Steamer, 6
  Tourist, 8
  Victor, 11

Bales-McCoin, Inc., 22
Ball disc, 41, 42-44
Battelle Columbus Laboratories, 22
Basis, 261
Beier, Joseph, 76
Beier variator, 77-81
Blanking press, 276-290
  flywheel design, 279
  motor selection, 278
  ratio range, 281
  starting transient, 283
Borg-Warner Corp., 22
Boring machine, 311
Briggs & Stratton, 25
British Leyland, 20
British thermal unit, 127
Brown, see David Brown Sadi

Capital, 261
  basis, 261
  equipment, 261
Cash flow, 263
  net cash flow, 266
Celsius, 140
Centigrade, 140
Chery, Walter, 20, 94, 101

Cleveland Gear, 28
Compactor, municipal waste, 290
Cone disc drives, 76
  Beier variator, 77
  planetary, 81, 84
Cone-on-cone stress, 182
Cone-on-cylinder stress, 182
Contact area analysis, 168
Contact stress, 178
Continuously Variable Transmission (cvt), 321
Contraves AG, 68
Cooling, heat exchanger, 302-303
Copy lathe, 310
Cost
  acquisition, 265
  depreciation, 266
  energy, 265-266
  maintenance, 263
  operating, 263
  opportunity, 271-272
Crack propagation failure, 186
Creep, 144, 169-175
Curtiss Wright Corp., 17
Curvature, radius of, 178
  principal radius of, 180
Cutoff machine, 303
Cylinder-on-cylinder stress, 182

David Brown Sadi, 87
Decanewton, 139
Depreciation, 261-263
  straight line, 262-263
Diameter
  mean rolling, 146
  parallel contact, 169
  pattern, 146
  transverse contact, 163-169
Dimensional analysis, 121
Discounted cash flow, 263
Disc drive, 76
  Beier variator cone, 77-81
  planetary cone (Lenze), 81-84
  planetary cone (Shimpo), 84-86
Drive specification sheet, 248
Dynamic viscosity, 135

Index

Efficiency, 130
Electric motors (AC)
  enclosures, 214
  flange and skirt mounted
    C flange (IEC), 214
    C flange (NEMA), 213
    D flange (IEC), 213
    skirt frame (IEC), 215
  frame descriptions
    single-phase motors, 209
    polyphase open type, 210
    polyphase totally enclosed, 211
  frame sizes, table, 208
  horsepowers, standard, 207-209
  IEC styles, 207
  metric styles, 207
  mounting, 208, 212-215
  NEMA styles, 207
  polyphase, 218
    also see torque speed characteristics
  rotor inertia, 229
  single-phase, 228-234
    also see torque speed characteristics
  torque speed characteristics, polyphase
    NEMA A, 225
    NEMA B, 221
    NEMA C, 226
    NEMA D, 226
    NEMA F, 227
  torque speed characteristics, single-phase/split-phase
    capacitor split, 232-233
    resistance split, 231-232
    shaded pole, 234
  torque values
    breakdown types A, B, C, 224
    locked-rotor types A-D, 222, 223
    pull-up types A, B, C, 223

Electric Rates, typical, by region, 271
Energy
  heat, 127
  kinetic, 124
  potential, 126
Energy factor, 165
  function of spin rate, 171, 172, 174-175
  and wear, 167
Erban, Richard, 15, 86
Eurodrive, Inc., 110
Excelermatic, Inc., 86, 106
Expense, 263

Factor
  energy, 166, 171-172, 174-175
  lubrication, 186
  pad, 166
  service, 237
  shock, 237
Fafnir Bearing Company, 20, 94, 101
Fire hazard, 296
Floyd Drives Company, 68
Flywheel, 279, 280
Fluids
  napthenic base, 155
  paraffin base, 155
  Santotrac, 19, 155
  synthetic traction, 155
Force, 120, 139
  contact, 145, 146
  tangential, 135
Free Ball drive, 68
  bidirectional four ball and roller, 70, 71
  four ball and roller, 71-73
  single ball, 73-75
  two ball and control roller, 75-76
Frequency, natural, 121
Friction drives, 109
  parallel disc and ring, 110-115
  perpendicular disc, 115
F.U. (Fabrication Unicum), 42

Garret Corp., 22
Gates Rubber Co., 22
General Motors Corp., 15, 16
  New Departure, 16
  Research, 15, 18
Graham, Lou, 25
Graham Transmissions, Inc., 42, 46
Grinding machine, 312

Hans Heynau GMBH, 32, 47, 68
Hayes, Frank, 15, 86
Headstock drive, 310, 312
Hertz stress, 177
Heynau, see Hans Heynau
History
  automotive, 5-24
  developments
    early, 6-19
    fluid, 19
    modern, 19-24
Hunt, Charles, 24, 86

Inertia, 127-129, 240-241, 254-255
  load, 228, 240
  moment of, 127
  normal, 244
  reflected, 128, 241
  rotor, 228
Interest
  annuities, 273
  compound, 272, 273
  equivalent, 274
  rate tables, 268-269
Irreversibility, 130

Jam protection, 291-295

Kelvin, 140
Kemper, Yves, 20
Kettering, Charles, 15
Kilogram, 138
Kilopascal, 139
Kilopond, 139

Kinematic viscosity, 138
Knitting machine, 296
Kopp, Jean E., 25, 27, 28
Kopp variators
  ball, 28-35
  roller, 38-41
  type M, 35-37
Koppers Company, 27, 28
Kraus, Charles E., 16, 19, 27, 86, 144 (note), 168 (note)
Kumm Industries, 22

Lathe headstock drive, 312
Length, 120, 138
Life, 186-204
  calculation, 187
  load spectrum, 193-196
  Poisson distribution, 196-197
  prediction method for designers, 186-187
  prediction method for users, 199-204
    ball-on-cone drive, 199-202
    cylindrical contact drive, 202-204
    toroidal drive, 199-202
    Weibull distribution, 196, 198-199
Loads
  aiding or overhauling, 236-237, 259
  characteristics of, 237-240
  constant horsepower, 236
  constant torque, 236
  cyclical torque, 237
  direction, 236-237
  fan law, 235
  inertia, 230, 240
  overhung, 249
  service factors (table), 238
  spectrum, 193-196, 202-204
Loewenthal, Stuart, 97
Loss, 165, 170, 172-173, 175
Lundberg-Palmgren theory, 187

Index

Machine tools
 copy lathe, 309
 boring machine, 311
 grinding machine, 312
 headstock drive, 312
 milling machine, 311
 stamping press, 311
 thread-cutting machine, 312
Maschinenfabrik Hans Lenze KG, 77
Mass, 120, 138
Mass moment of inertia, 129
McDonough Power Equipment Company, 110
Mechanical stops, 130
Meter, 138
Metric horsepower, 140
Metric system, 138-143
 conversions (table), 142, 143
 force, 139
 power, 140, 141 (table)
 pressure, 139, 140 (table)
 temperature, 140, 141 (table)
 torque, 139, 140 (table)
Metron Instruments, Inc., 87
Milling machine, 311
Mixer, 298
Monsanto Corporation, 19, 155
Moment of inertia, 127
 of mass, 129
 of important solids, 129
Motion, skew, 145
Motors, see Electric motors

Nastec, Inc., 94, 97
Nasvytis, Algirdis, 27, 99
National Aeronautics and Space Administration, 27, 97
Net present value, 263
 acceptance criteria, 271
Newton, Isaac, 120
Newton (force), 139

Offset sphere drive, 109
Opportunity cost, 271
Overhung load, 249-250

Paint mixer, 298
Paxton Products, 94, 101
Perbury, 16, 20
P.I.V. Antrieb Werner Reimers KG, 42
Planetary drives, 94
 constant ratio ball, 101
 constant ratio roller, 97-101
 constant ratio wedge roller, 95-97
 variable ratio, 101-106
  externally adjusted, 103-104
  internally self-adjusting, 105-106
Pouseille, J. L. M., 136
Poisson distribution, 196
Pound (mass), 121
Power, 124, 140
 recirculating, 131-135
Pratt and Whitney, 17
Precision (speed setting), 257
Pressure, 139
 contact, 162
 profile, 147

Radius
 of curvature, 178
 of gyration, 127
 of important solids (table), 129
Rate
 fluid film shear, 165
 mean rolling, 165
 spin, 169-174
Recirculating power, 131-135
Ring cone drives, 46
 dual sun wheel, 55-58
 nutating cone and ring, 63-68
 parallel, 58-60
 with planetary, 48-51
 single-sun wheel, 51-55
 variable-pitch diameter, 62-63
 variable-position input shaft, 60-62
Roller disc drives, 41, 44-56

Rolling Traction Analysis and Design, 144, 168
Rouverol, William, 41

Sadi, see David Brown Sadi
Shear, 144
  geometry produced, 145
  motion, 144
  rate, 150, 162, 163
Shimpo Industrial Co., Ltd., 46, 47
Shock, 256
  factors, 237
Side load, 247
  see also Overhung load
Slip, 145, 221
Sloan, Alfred, 15
Slug, 121
Solar energy plant, 307
Spalling, 186
Specific heat, 127
  table of, 128
Specification sheet, 246-260
  access, 259
  adjustment items, 252-254
  input items, 247-251
  load items, 254-259
  mounting items, 254
  operating conditions, 259
  output items, 251-252
  power source, 259-260
Speed, 235
  changes and reversals per hour, 256
  holding precision, 258
  input, 247
  output, 251
  range, 251
  ratio, 164
  rolling, 151
  synchronous, 218
Speed adjustment
  at rest, 258
  repeatability, 258
  through zero, 258
Speed-up drives, 303-305

Sphere-on-cone stress, 178
Sphere-on-toroid stress, 183
Spin, 145, 159-165, 168
Spool drive, 106
Stamping press, 311
Starting transient, 283-285
Stress, 186
  cone-on-cone, 182
  cylinder-on-cone, 182
  cylinder-on-cylinder, 182
  footprint, 177
  hertz, 147, 177-186
  mean contact, 152
  sphere-on-cone, 178
  sphere-on-toroid, 183
Sumitomo Machinery Corporation of America, 77

Temperature, 140, 142-143, 153
Thread-cutting machine, 312
Toroidal drives, 86
  dual cavity, 92-94
  with planetary, 89
  single-cavity, 92-94
    single-cavity two-roller, 87-89
    single-cavity three-roller, 89-92
Torque, 122, 130, 139, 235, 247, 252, 255
  also see Electric motors
Traction
  angle, 149
  coefficient, 146-154, 169, 171-175
    tables, 159
  curve, 150
  fluids, 154
Traction drive, defined, 144
Traction Propulsion, Inc., 17, 106
Tracor, Inc., 17
Transitorq, 16, 24
Transmission
  automatic, 15
  friction disc, 6
  synchromesh, 15

Vadetec Corp., 20, 47
Value
  net present, 263-264,
    271-272
  scrap, 262-263
Van Doorne Transmissio, 22
Velocity, rolling, 164
Viscosity, 135-138
  change during processing,
    298-310
  dynamic, 135
    table, 137
  index, 154-155
  kinematic, 138

Voltages, standard in Europe
  and U.S., 260

Waste compactor, 290
Watt, 140
Wear, 153
Wedgetrac Corp., 94, 95
Weibull distribution, 198
  function paper, 201
  tables, 200
Winsmith Corp., 28
Work, 123

Young, J. R. Corp. 109